Proceedings of the CEC International Symposium / Ispra
4-6 March 1986

INTEGRATED TSE-TSE FLY CONTROL: METHODS AND STRATEGIES

Edited by
R.CAVALLORO
Commission of the European Communities, Joint Research Centre, Ispra

Published for the Commission of the European Communities by
A.A.BALKEMA / ROTTERDAM / BROOKFIELD / 1987

*The texts of the various papers in this volume were set individually
by typists under the supervision of each of the authors concerned.*

Publication arrangements: *P.P.Rotondó*, Commission of the European Communities,
Directorate-General Telecommunications, Information Industries and Innovation, Luxembourg

EUR 10530

Published by
A.A.Balkema, P.O.Box 1675, 3000 BR Rotterdam, Netherlands
A.A.Balkema Publishers, Old Post Road, Brookfield, VT 05036, USA

ISBN 90 6191 702 6

ORGANIZING COMMITTEE

CEC General Directorate Science, Research and Development
    DG XII - Div. A.4

    General Directorate Development
    DG VIII - Div. A.2 and Div. 5

RESPONSIBLE

    R. Cavalloro - Principal Scientific Officer CEC
    DG XII - Joint Research Centre - Ispra

SECRETARY

    M.P. Moretti - Public Relations and Presse
    DG XII - Joint Research Centre - Ispra

SESSIONS' STRUCTURE

	Chairman	Key-Speakers
Opening session: Opening address and introduction	D.H. Davies	A.M. Jordan
Session 1: The safety and efficaciousness of chemical control	S. Geerts	I.J. Everts
Session 2: Biological, biotechnical and other control methods	J. Mutuku Mutinga	D.A. Dame
Session 3: Present and future programmes of international and national organizations	J. Mulder	K.B.David-West
Session 4: General aspects on Glossina	C. Pelerents	
Closing Session: Sessions' report and conlcusions	C. Pelerents	A.M. Jordan I.J. Everts A. Dame K.B.David-West

Commission of the European Communities - Joint Research Centre, Ispra
Conference-hall N. 3

Foreword

Despite the considerable and praiseworthy efforts conducted by many, and the successes which have certainly been obtained up to now, the serious problem of sleeping sickness, unfortunately, is still present and preoccupying. In all the Countries where this protozoarian sickness is endemic, in particular in a good 36 African Countries, its control has become necessary and urgent.

After about a century from the discovery of the Diptera of the genus Glossina as agents of serious pathogenic trypanosomiasis, and despite the immediate recourse to extensively used chemical methods, the incidence of this sickness remains vast and the problem is of great actuality.

The aleatority of appropriate prophylactic methods and of the use of pharmacological products, as well as the appearance of chemical-resistant stocks, justify the great recourse to the direct control of ematophageous vector flies, which are the main active transmitters of the sickness. Their key role is becoming increasingly obvious.

As is well known, disease transmission limits livestock development and human health and has a very great economic importance in the intertropical areas of Africa.

The Commission of the European Communities, which is involved in research and development of the control of trypanosomiasis in Africa, felt that it would be extremely useful to give a valid contribution to the solution of the pressing problem by organising a specific symposium, with the intention of making more rapid progress in the struggle against Glossinae. The accent has mainly been put on the validity of the recourse to integrated control.

The meeting allowed a glimpse of a better application of the results already acquired, a comparison of the various methods for control of the Tse-tse fly and a more suitable definition of the strategies which can be used in the various environments of interest.

This publication collects all the papers given at the symposium, in which eminent experts from 12 countries and 4 international organisations took part.

As well as specific themes, which considered the safety and efficaciousness of chemicals, their ecological impact, the biological methods, the sterile insect technique and specific biotechniques which recourse to insecticide-treated screens or targets, traps, odour baits, etc., a wide survey of the lines of action under way was given and future programmes planned by the national and international organisations interested in the problem were presented.

The exchange of many and fruitful ideas and the thorough discussions stressed the need to develop joint actions and to increase coordination and cooperation in the struggle against this scourge, which is still today dramatically alive and a worry.

R. Cavalloro

Contents

Opening session

Session 1. *The safety and efficaciousness of chemical control*

Session 2. *Biological, biotechnical and other control methods*

Session 3. *Present and future programmes of international and national organizations*

Session 4. *General aspects on Glossina*

Closing session

Opening session

Chairman: D.H.Davies

Opening address

D.H.Davies
Directorate-General Energy, CEC

Let me begin by extending to you a most sincere welcome on behalf of the Commission of the European Communities. The presence of leading experts at our Symposium is for us a guarantee of success and I want to thank you most sincerely for your participation.

Tsetse flies play today, as they always have, an unquestionably important and harmful role as a vector for disease, both of man and of livestock. The damage regularly suffered by a large number of African countries from this cause is very large.

The discovery by Bruce of diptera of the genus <u>Glossina</u> as agents of severe pathogenic trypanosomiasis dates back a long way — as we know — to 1895; some years later, in 1905, Castellani found infectious micro-organisms in the cerebrospinal fluid of patients suffering from sleeping sickness.

Today, nearly a century later, trypanosomiasis is still a scourge which is far from being conquered despite the magnitude, and the success, of the efforts made up to now.

The use of new chemotherapeutic products has, without doubt, given positive results but the secondary toxic effects, the appearance of strains of chemoresistant trypanosomes, their capacity to modify their structures to resist the actions of their host's immune system, and the difficulties of taking appropriate prophylactic measures all concentrate efforts towards the control of the tsetse fly, which is such a very active transmitter of disease.

In practice, the epidemiology of this protozoal disease is closely linked to the behaviour of the bloodsucking vector flies, wild species which play an important economic role in this context, by reason of their influence on conditions of life in vast agricultural and pastoral regions.

To gain an impression of its importance, it is enough to remember that an effective control of tsetse flies could make possible developments in inter-tropical Africa which would be of exceptional nutritional importance on world scale. An additional seven million square kilometers could be gained for agriculture and stock breeding, a greater land area than that used for agriculture in the whole United States of America.

An interesting evaluation by the Food and Agriculture Organization (FAO) shows that once trypanosomiasis has been overcome, the total savannah pasturage infested by tsetse flies could support a livestock level of 120 million cattle, making available around one and a half million tonnes of meat every year; and we all know the importance of the provision of protein in a world where the food crisis and the problem of famine become ever more serious.

The campaign against trypanosomiasis has taken several forms: detection of flagellated protozoa in an organism and then intervention with pharmacological products, vaccination, selection of stock resistant

3

to infestation and, finally, direct action against the trypanosome vector, the tsetse fly itself; these different approaches are all the object of simultaneous research and its application, and none must be neglected.

Many worthwhile initiatives have been undertaken on an international level. These have led to a very careful study of pathogenic strains using controlled stock-raising in laboratories on semi-artificial substrates or in the field (dealing with ecological aspects and population dynamics, sexual reproduction and behaviour, nutritional needs and the process of digestion, the importance of the role of hormones, etc.), and to the use in this campaign of very varied methods of intervention such as agroforestrial, physical, chemical, biological, biotechnical, genetic, etc.).

Secondary effects which are often harmful for the environment, follow the large-scale use of insecticides, as does the appearance of the phenomenon of resistance. These effects stimulate a more considered and rational utilization of these products, and also the introduction of means of action more specific and less harmful to the balance of nature in the local settings under consideration.

Possible research ideas on this general topic are numerous and come from many different sources, leading to activities involving numerous institutions including the international organizations which are actively engaged in this work. I should emphasize, amongst others, the efforts made by the World Health Organization, and especially the Special Programme for Research and Training in Tropical Diseases organized jointly by the World Health Organization, the United Nations Development Programme and the World Bank, with its Scientific Working Groups on African Trypanosomiasis and on Biological Control of Vectors, as well as the work of the International Atomic Energy Agency and FAO on the application of sterile insect techniques. Let us remember too the work of the Institut d'Etudes et de Medécine Vétérinaire Tropicales and the Tropical and Development Research Institute on attractants and traps, as well as that of the Institut Français de Recherche Scientific pour le Développement en Coopération and the Deutsche Gesellschaft für Technische Zusammenarbeit on this subject.

The Commission of the European Communities is very much aware of this whole problem and wishes to make an effective contribution to the fight against trypanosomiasis : it has already initiated activities within the framework of its cooperative actions with developing countries.

Two Directorates-General, those for Development and for Science, Research and Development, are concerned with this problem. Within the activities under the Lomé Convention between the Commission and 65 Third World countries, of which the majority are in Africa, a vast regional project has been started for the control of tsetse flies over several millions of square kilometers in Zimbabwe, Zambia, Mali and Mozambique. The emphasis here has been put mainly on the long-term use of non-chemical methods with the aim of permitting the opening, or sometimes the reopening after years of war, of vast zones to agriculture. Alongside such projects for vector control, others are endeavouring in West Africa to achieve improvement in herd quality by the introduction of trypanotolerant animals. Within its activities in scientitific research, the Commission has brought into being a programme entitled "Science and Technology for Development", with one subprogramme dealing with Tropical Agriculture and the other with Tropical Medicine. The majority of research programmes supported by the Commission involve cooperation between European organizations and those in the Third World, dealing particularly with subjects which concern the latter.

In Medicine as much as in Agriculture, the question of trypanosomiasis occupies an important place, notably in the campaign against such diseases by using the expedient of vector control. The actual research projects put an emphasis on the use of efficient and specific attractants and traps, mainly in the Ivory Coast and Burkina Faso. In parallel with this approach to the problem, important research is also taking place on the natural resistance to parasites of certain races or individuals in various subsaharan African countries. In future, this programme will always consider trypanosomiasis as a priority and it can act as the catalyst for a research network which will define and coordinate the work necessary for the introduction of an integrated control strategy against the tsetse fly.

Before finishing I would like to emphasize one problem, lying downstream from the fight against the tsetse flies themselves and which is being looked at with much attention by the Commission, namely that of the appropriate use of land when returned to agriculture. Careful planning of the livestock level to take account of the capacity of the local environment will prove to be absolutely indispensable if we are nor to introduce a new scourge which is that of overgrazing and the resulting desertification.

The work which is going to take place here at Ispra will indoubtedly lead to a better knowledge of current research, the developments being carried out at field level, the results obtained and the problems which remain to be solved.

The great interest aroused by this Symposium is thus well justified, and your presence here today is a most positive proof of this.

May I wish you all a hardworking and worthwhile meeting which will, I am sure, accelerate progress towards the control of the tsetse fly, an insect vector which is sadly all too well-known and which unfortunately remains all too topical a challenge.

Prospects for integrated control of the African trypanosomiasis

A.M.Jordan

Tsetse Research Laboratory, ODA/University of Bristol, Langford, UK

Summary

Trypanocidal drugs are the main agents used for control of African
human and animal trypanosomiasis and it is argued that it is
therefore more appropriate to consider strategies for integrated
disease control rather than integrated control of the vector, the
tsetse fly, alone. Eradication of animal trypanosomiasis within a
prescribed area, which has been achieved in the past, is only
possible following removal of the vector which can in turn only be
achieved, except in the rare circumstances of small isolated
infestations, by dealing with large areas by a roll–up–the–country
approach. Expensive campaigns can usually be justified as the
objective should be to treat places once only and prevent reinvasion
by natural isolation or development of the land involving removal of
fly habitats and hosts. Eradication of the human disease alone by
vector elimination is not a realistic objective. Control of both
animal and human disease implies regular operations and, except under
exceptional circumstances, expensive methods of vector control are
inappropriate. The strategy should be, wherever possible, to combine
the controlled use of trypanocidal drugs with low cost methods of
vector control. Methods which might appropriately be employed under
these various circumstances are discussed.

1. Introduction

It should not be forgotten that tsetse flies are harmless insects and
it is the trypanosomes that they carry which are pathogenic to man and to
his domestic livestock. In all African countries where trypanosomiasis is
endemic some attempt to control the effects of these parasites is made by
the use of trypanocidal drugs – more effectively in some countries than in
others – whereas only in a few African countries is any attempt made to
control the vectors of trypanosomiasis. This is only one reason why one
objective of this paper is to attempt to broaden the scope of this
symposium by considering prospects for integrated control of the African
trypanosomiases, in which vector control is envisaged as playing a key
role, rather than integrated vector control alone. Although this approach
acknowledges the present prominence of chemotherapy for control of
trypanosomiasis in Africa, I believe it also provides the most realistic
way forward for the future.

The integrated control of insect pests is very much a modern concept
but one, as far as tsetse flies are concerned, to which so far little more

than lip service has been paid. Most tsetse control campaigns have depended almost entirely on one control method with perhaps limited back-up by one or more other techniques. A fundamental limiting factor is that the larval stages of _Glossina_ – the most vulnerable life cycle stages in many insect pests – are spent almost entirely within the female fly and thus are not available as targets for control. Because of the limited options therefore available, this again emphasises the importance of considering vector control techniques as components of an integrated approach to disease control.

So far reference has been made rather loosely to tsetse and trypanosomiasis control, implying the reduction of the incidence of vectors and of the disease but the continued presence of both. However, complete eradication of both vectors and this disease within prescribed areas has been achieved in the past and it is convenient to retain the distinction between the objectives of control or eradication, as optimum strategies in the two different sets of circumstances are different. This has not always been appreciated in the past but is becoming increasingly clear, especially now that economic factors, particularly cost/benefit analyses are necessarily being taken into account before the initiation of operations. There are also clear differences between strategies appropriate for combating human and animal trypanosmiasis.

2.   Methods of trypanosomiasis control or eradication

There have been many reviews (6) of past and present methods of trypanosmiasis control which it is inappropriate to repeat here. Table I lists those methods which are worthy of consideration as components of

TABLE   I

Methods	Remarks
A. Suitable for possible inclusion in integrated strategies of disease control	
a) Directed against the parasite:	
Trypanocidal drugs	Curative and prophylactic
b) Directed against the vector:	
Clearing of vegetation	Includes pasture management to control extent of tsetse habitat
Appln. of residual insecticide	From the ground and from the air
Appln. of non-residual insecticide	Especially from the air
Insecticide-impregnated targets and traps	
Sterile insect technique	
B. Unsuitable for integrated strategies of disease control	
a) Directed against the parasite:	
Immunization	Remote possibility for the future
b) Directed against the vector:	
Destruction of wild animal hosts	Unacceptable
Biological control)	No practical method yet
Physiological control)	developed

8

integrated strategies for the control of trypanosomiasis. Other methods
are also listed which, for a variety of reasons, appear to be unsuitable
for inclusion in such strategies. Still other approaches, such as the
enforced removal of people and domestic animals from contact with tsetse,
and the essentially passive approach to control of the animal disease of
keeping breeds of cattle tolerant of its effects, are considered
inappropriate for discussion in the present context.

3.   The eradication of animal trypanosomiasis

   The only way in which in the past the threat of trypanosomiasis has
been removed by man from a prescribed area has been following elimination
of the local tsetse vectors and this is still the only realistic approach
to eradication of the disease. Under such circumstances integrated anti-
vector methods, with no role for trypanocidal drugs other than as a stop-
gap until the objective of eradication is achieved, is the appropriate
strategy. Many past campaigns in a number of countries have shown that it
can be relatively easy, subject to the competent execution of appropriate
measures, to eliminate tsetse from an area but much more difficult to
ensure that the area remains free of flies.
   Experience has shown that to prevent reinvasion of areas cleared of
tsetse either the area has to be relatively small and well-isolated from
other infestations, or large and treated on a roll-up-the-country basis
with consolidation of cleared areas soon followed by settlement and
development. Very few tsetse belts are small and isolated, but successful
campaigns against G. palpalis were conducted on the Island of Príncipe
(126 km^2) by Da Costa et al. (1) and again, following reintroduction of
the fly by De Azevedo et al. (2). The Zululand belt of G. pallidipes
(some 18,000 km^2) was also well-isolated from tsetse elsewhere and was
removed by Du Toit (3) in the first major operation employing
insecticides.
   Successful large-area eradication campaigns have also been limited in
number. Perhaps the most spectacular has been the removal of Glossina
species from some 200,000 km^2 of northern Nigeria where consolidation
depended on people and their stock moving into the cleared areas and
taking over the land, removing habitats and hosts of the flies in the
process. In Uganda this degree of land hunger did not exist and when many
years of efficient anti-tsetse operations (10) came to an end, no
defendable perimeter had been reached, insufficient people moved on to the
cleared land and tsetse reinvaded. In Zimbabwe painstaking and effective
work over many years was lost when tsetse reinvaded cleared areas when
anti-tsetse operations had to stop during the war preceding independence.
The same lesson has been learned elsewhere.
   What are the prospects for the future? Small isolated infestations
of Glossina do occur - on Zanzibar island for example - but these are
uncommon and the economic case for removing tsetse is not established and
should be argued in each set of circumstances. However, these special
circumstances are rare and when considering tsetse, and hence disease,
eradication, it must be with large areas, and with the subsequent
development of these areas, with which we must primarily be concerned.
Today the only effective large-area campaign in progress is in Zimbabwe,
which it is intended will soon be linked with the recently initiated
Regional Programme involving the planned eradication of tsetse from some
320,000 km^2 of Malawi, Mozambique, Zambia and Zimbabwe. The possibility
for other large area campaigns exists, for example other Nigerian-type
campaigns starting at the northern limit of fly in West Africa and working

9

southwards with land freed of fly being consolidated by settlement of people and development of the land. There are similar prospects elsewhere around the fringes of tsetse-infested land in Africa, but it is unrealistic for such large area operations to be initiated in countries other than those at the edges of the infested area or where no readily defendable natural perimeters of the infestations exist. As tsetse do not recognise international boundaries, usually regional operations will be necessary. Whether or not other large-area operations will become possible will depend on many factors which need not concern us here but which include environmental, economic and political considerations and the availability of funds from international sources.

Eradication operations in the past have mainly employed one technique for killing the flies. All successful large-area campaigns have depended almost entirely on the application of residual insecticides applied by men from the ground equipped with some type of sprayer. In Nigeria these operations were backed up with similar applications using helicopters but only some 5% of the total area sprayed was treated in this way. Ground spraying has been effective in most types of savannah vegetation and terrain and is certainly still the most reliable single method of tsetse eradication. However, the method is labour-intensive and, to be successful, the large teams of men have to be organized on almost military lines. Such logistic problems, financial pressures, and the pressure to stop using the relatively inexpensive chlorinated insecticides such as DDT (despite highly specific use in anti-tsetse operations) have resulted in the run-down of ground spraying activities in recent years.

No single technique has evolved to take its place - hence, partly, the interest in planning campaigns in which are combined a number of different methods. If large areas are to be covered there is, at present, no doubt that the main method must involve the use of insecticides. The aerial application of non-residual inecticides from fixed wing aircraft is now the most popular approach to pesticide application. It can be virtually non-polluting, can cover large areas and, perhaps most attractive, it can be contracted out to a private operator thus doing away with many of the problems of organizing large groups of men on the ground. Nevertheless the method cannot, so far, achieve eradication under all conditions of climate and terrain and does not obviate the necessity for competent ground-based teams to undertake pre- and post-spray surveys and to monitor the activities of the contractor.

In what way may other techniques be integrated with this basic method? One problem is the existence of remnant pockets of tsetse in an otherwise cleared area. These may be attacked by the application of residual insecticides - either from the ground or from the air - but first it is necessary to demonstrate their existence, not easy when large areas are being treated. Insecticide-impregnated targets, with associated odour attractants in the case of the savannah species, could also have a role to play both in the mopping up of residual foci and the treatment of areas where aerial spraying is of doubtful efficacy, such as deep, narrow gullies. They could also, it seems (Vale, personal communication), possibly serve as a temporary barrier to prevent flies entering a cleared area from a neighbouring area, although it must be emphasised that no man-made barrier has ever been effective for more than a short time and there is no effective substitute for resuming roll-up-the-country eradication operations until an extensive natural barrier to fly movement is eventually reached.

The sterile insect technique has also been seen by some as a valuable component of integrated approaches to the eradication of tsetse.

Certainly it is an elegant, specific, non-contaminating approach, most efficient when fly densities are low (when they are costly to attack by insecticides) and one whose effectiveness has been demonstrated in small-scale experiments (7 & 9). However, there are disadvantages, especially the sophisticated, expensive facilites and skilled manpower that would be needed to produce the enormous numbers of male insects, generally of more than one species, for sterilisation and release before the technique could be incorporated into large-area eradication campaigns. Further, it has been argued that if other methods can achieve eradication why stop before this objective is achieved and change to another method? At the end of the day decisions on whether or not to include the technique in large-area eradication operations will depend both on its feasibility and on its cost compared with other methods.

As we have seen, an important contributory factor to the success of the Nigerian campaign was the speed with which land cleared of tsetse was settled by people who removed fly habitats and hosts. Whereas the deliberate removal of vegetation as a means of tsetse control, widely practised in the past, is unlikely to be resurrected in the future, it should be emphasised that bush control and pasture development and management are amongst the keys to improved systems of traditional livestock husbandry and should be seen as components of integrated approaches to the eradication of Glossina. By definition, eradication does not mean simply killing all the flies in a prescribed area; it also means ensuring that the area remains free of fly.

4.   The control of animal trypanosomiasis

Whereas eradication of trypanosomiasis has only been achieved following disappearance or removal of the vector, control of the disease has been, and still is, primarily effected by trypanocidal drugs. Although eradication may be very expensive to achieve, if successful it has many economic advantages over a continuing commitment to control. If insecticides are employed it also has environmental advantages as regularly repeated applications are avoided.

The main role of trypanocidal drugs is to treat or to protect cattle from trypanosomiasis in areas where the numbers of tsetse are relatively low and the trypanosomiasis challenge thus presented to domestic livestock is light. Although it is possible to protect cattle with drugs even in the presence of dense infestations of Glossina (8) this can only be achieved when the standards of management and veterinary attention are far higher than those available to most of the traditional livestock owners of Africa. However, there are many problems associated with the use of drugs (availability, extent of veterinary support, drug misuse, drug-resistance, etc.) which need not concern us here but are so great that a strong case can be made for tsetse control as a supporting strategy.

But first the limitations of tsetse control must be clearly appreciated. In the past, and even sometimes today, the objectives of tsetse eradication and tsetse control have been confused. For instance, it is possible to "eradicate" tsetse from an enclave within a sea of infestation. The Shika Stock Farm was maintained free of tsetse for several years in the 1960s by the annual ground spraying of residual insecticide (5). In 1968 it was estimated that the annual cost of this was about £13,000 compared to the, then, value of the 360 cattle on the farm of about £10,800 (4); spraying was stopped and tsetse soon become re-established. In more recent years the aerial spraying of enclaves within major fly-belts has been equally uneconomic.

Whereas to achieve the objective of eradication it is usually necessarily to "think big", the objective of control, to be economically viable, must generally involve the use of low cost methods. Although drugs will undoubtedly continue to be the main weapon, there are realistic prospects for combining this with cheaper methods of tsetse control – not to kill the last fly but to reduce fly populations such that the level of challenge can realistically be tackled with drugs. The opening up of new areas of pasturage in this way can be envisaged.

Under these circumstances it seems likely that the use of insecticides, costly to apply, purely for purposes of tsetse control will become much more difficult to justify in future. However, there are circumstances where their use can be justified – for instance, when it is only necessary to spray regularly a small area in order to prevent the invasion by <u>Glossina</u> of a large naturally fly-free area or the reinvasion of areas from which tsetse have been removed.

Other costly methods, such as the sterile insect technique, are equally inapplicable to meet a commitment for repeated control of <u>Glossina</u>. Insecticide-impregnated targets and associated odour baits may be cheap enough to operate continuously, but even these devices have to be serviced regularly and the costs involved have yet to be worked out. The savannah species of <u>Glossina</u>, the principal vectors of animal trypanosomiasis, are widely dispersed and the logistics and costs of locating and servicing a significant number of targets would not be inconsiderable, but could be acceptable if balanced against the value of new land that could be utilized by domestic livestock with limited use of trypanocidal drugs.

5. The eradication of human trypanosomiasis

The eradication of sleeping sickness within a prescribed area is generally not a realistic objective unless foci of the disease are within the much wider area covered by a campaign to eliminate the vectors of animal trypanosomiasis. A number of sleeping sickness foci disappeared in this way following the tsetse eradication operations in northern Nigeria which were primarily aimed at the removal of animal trypanosomiasis.

6. The control of human trypanosomiasis

Vector control is not a cost-effective option for the control of endemic sleeping sickness over large areas and control of the disease under these circumstances will have to continue to rely on the ability of rural medical services to diagnose and treat the disease. On the other hand, during epidemics, which are usually fairly circumscribed in area, vector control becomes an acceptable strategy. However, as with the animal disease, it is desirable to integrate vector control measures with surveillance of the human population and the use of trypanocidal drugs to treat diagnosed cases. Again one must decide what vector control methods should be applied. Financial considerations should not be ignored but, quite rightly, strict cost/benefit analyses are not called for as money can generally be found to deal with an outbreak of human disease. The choice of technique depends on the resources available and on the ecology of the flies; speed is generally essential. Insecticide methods can be used, but simple insecticide-impregnated screens offer a cheap non-polluting alternative and with the linearly-distributed vectors of <u>T. b. gambiense</u> sleeping sickness, odour attractants are not a necessity. The objective should be to reduce the number of vectors so that, combined with a reduction of the mammalian reservoir of the disease by surveillance and treatment of cases, transmission of the disease is significantly reduced.

The sterile insect technique is not amongst the options for inclusion in vector control operations as even a temporary man-induced increase in the number of potential vectors of sleeping sickness would be unacceptable.

7.   Conclusions

Before embarking on campaigns to control African trypanosomiasis it is necessary to define objectives - is control really the objective or is it possible that eradication can be achieved and sustained? Is human or animal trypanosomiasis being combatted? A rational choice of the most appropriate anti-vector or anti-disease techniques can then be made. However, even if a logical choice of methods results from such an assessment, final decisions are frequently influenced more by logistic and economic than purely technical considerations. As we have seen, ground spraying of residual insecticide is, under many circumstances, the method of choice but is becoming impossible to manage. A regrettable fact of life.

The concept of integrated tsetse control, and even the wider one of integrated trypanosomiasis control, is still somewhat idealistic. Nevertheless, techniques do exist by which the effects of these diseases can be reduced and the main limitations to their more effective deployment are essentially logistic and financial. Certainly it would be helpful if research could indicate new, preferably cheap, methods of control (automatic traps for sterilising both sexes of tsetse, biological control agents, vaccines), but this is not the main problem. Combinations of existing methods, competently applied, can achieve effective control of African trypanosomiasis, in many circumstances at an acceptable cost. It is mainly a question of choosing the right methods, organization and effective man-management.

REFERENCES

1.   **Da Costa, B.F.B., Sant'ana, J.F., Santos, A.C.** and **Alvares, M.G.A.** (1916) Sleeping Sickness. A Record of Four Years War Against it in Principe, Portuguese West Africa. Bailliere, Tindall and Cox, London.
2.   **De Azevedo, J.F., Da Costa Mourao, M.** and **De Castro Salazar, J.M.** (1962) The Eradication of Glossina palpalis palpalis from Príncipe Island. Junta de Investigacoes do Ultramar, Lisboa.
3.   **Du Toit, R.** (1954) Trypanosomiasis in Zululand and the control of tsetse flies by chemical means. Onderstepoort J. Vet. Res. 26, 317-87.
4.   **Jordan, A.M., McIntyre, W.I., Le Roux, J.G.** and **Negrin, M.** (1978) Economic assessment of present operations for the control of animal trypanosomiasis in Nigeria. Unpublished report. Food and Agriculture Organization of the United Nations, Rome.
5.   **Kirkby, W.W.** (1963) A review of the trypanosomiasis problem on Shika Stock Farm from 1929 to the present day. Bull. epiz. Dis. Afr. 11, 391-401.
6.   **Mulligan, H.W.** (ed.) (1970) The African Trypanosomiases. George Allen and Unwin, London.
7.   **Politzar, H.** and **Cuisance, D.** (1982) SIT in the control and eradication of Glossina palpalis gambiensis, pp. 101-9 in Proc. Symposium on Sterile Insect Technique and Radiation in Insect

    Control. International Atomic Energy Agency, Vienna.
8. **Trail, J.C.M., Sones, K., Jibbo, J.M.C., Durkin, J., Light, D.E.** and
    Murray, M. (1985) Productivity of Boran cattle maintained by
    chemoprophylaxis under trypanosomiasis risk. ILCA Research Report
    No.9, Addis Ababa.
9. **Williamson, D.L., Dame, D.A., Gates, D.B., Cobb, P.E., Bakuli, B.** and
    Warner, P.V. (1983) Integration of insect sterility and
    insecticides for control of Glossina morsitans morsitans Westwood
    (Diptera: Glossinidae) in Tanzania. V. The impact of sequential
    releases of sterilised tsetse flies. Bull. ent. Res. 73, 391-404.
10. **Wooff, W.R.** (1966) Consolidation in tsetse reclamation, pp. 141-8 in
    International Council for Trypanosomiasis Research, Eleventh
    Meeting, Nairobi. OAU/STRC.

Session 1
The safety and efficaciousness of chemical control

Chairman: S.Geerts

Is resistance developing in tsetse flies? Susceptibility to two chlorinated hydrocarbon insecticides in sprayed and unsprayed populations of *Glossina pallidipes* in Kenya

D.A.Turner & T.K.Golder
International Centre of Insect Physiology and Ecology, Nairobi, Kenya

Summary

An investigation into the possible development of insecticide resistance in tsetse flies was carried out by comparing the suscepti- bility to topical applications of endosulfan and dieldrin between two populations of <u>Glossina pallidipes</u> Austen in Kenya; one from the Lambwe Valley, where there has been a long history of spraying operations, and the other from Nkruman which has never been sprayed. Dosage-mortality regression lines and LD data showed that wild-caught, non-teneral males and females from Lambwe were significantly more tolerant of both insect- icides than their counterparts from Nkruman, with the exception of males dosed with dieldrin. The differences were insufficiently great, however, to be incontestable evidence for the development of resistance in the Lambwe strain, although there were no obvious sources of variability between strains adequate to account for the differences in susceptibility.

Introduction

Given the genetic potential, which seems to be a feature of all natural populations, the development of insecticide resistance is primarily a matter of sufficient selection pressure over a sufficient period. Selection pressure is applied through the frequency of application and the dosage of insecticides, and is conditioned by population density and distribution, reproductive capacity and movement capability of the insect. From a consideration of some of these factors as they apply to tsetse flies Burnett (9) concluded that resistance might develop but only very rarely. The continuing high levels of kill obtained in control operations would appear to bear this out, and where there have been failures in eradication attempts, these have mostly been accounted for by re-invasion and/or the survival of residual populations resulting from enhanced tolerance of insecticide by pregnant females (1, for review). The fact remains, however, that no resistance studies have ever been carried out, presumably because there has been no compelling reason to suspect it.

The subject has not been altogether neglected, nevertheless. In the process of insecticide toxicity trials and studies aimed at providing practical guidelines on dosages and formulations (1, for review), a body of comparative susceptibility data has been acquired which could serve as base-line information for resistance studies. More recently, an investi- gation by computer simulation predicted that there were circumstances in which high frequencies of resistant genes could be attained in tsetse populations subjected to fairly typical spraying regimes (18). Biochemical studies have also shown that tsetse can metabolise DDT (18) and the dieldrin analogue, HEOM (4), by reductive dechlorination into less toxic

17

forms, indicating a potential for resistance in two insecticides commonly used for tsetse control.

Insofar as insecticidal pressure and population isolation are concerned in promoting resistance, the influence of these factors has been expressed, perhaps more intensely than elsewhere, on a population of Glossina pallidipes Austen inhabitating the Lambwe Valley in western Kenya. This population was therefore considered to be a prime candidate for an investigation of possible development of resistance, and the justification for this is enlarged upon in the following description of the history of spraying operations and other relevant factors. An additional justification was the suspicion of control workers that one of the insecticides being used in the Lambwe Valley, namely dieldrin, was giving less than the desired effect, though a number of other factors could also be responsible.

As tsetse fly-belts go, the G. pallidipes infestation in the Lambwe Valley is small in area (approx. 300 km^2) and isolated (the nearest other infestation being 70 km away), but in parts contains one of the densest fly populations ever recorded (2). Whereas livestock trypanosomiases have always been present, Rhodesian sleeping sickness also became endemic in the area in the early 1960's, and from 1968 to the present, various measures, mainly insecticidal, have been carried out, initially to contain the diseases and later to eliminate them through vector eradication. During the late sixties and early seventies tsetse control was aimed at confining the fly to its principal habitat of thicket within the Lambwe Valley Game Reserve in the southern half of the valley, by clearance of lowland thicket in adjacent settlement areas and ground spraying in the form of a series of four monthly applications of residual dieldrin (Dieldrex 15T 1.8% e.c.) to thickets on hillsides flanking the valley. Also in the period 1968-1971 a series of five aerial spray trials, employing first helicopters then fixed-wing aircraft, were conducted in the valley thicket (17, 2). This thicket consists of fairly discrete large blocks, and each was either blanked sprayed or strip sprayed on one occasion or another with residual deposits of dieldrin invert emulsion, at dosages which varied between 54 and 200 g a.i./ha. Post-spray monitoring recorded reductions in the G. pallidipes population of 95-99.9%, which were maintained for six to eight months afterwards.

Soon after the completion of the aerial spray trials the indications were that the tsetse population was rapidly recovering (3). Because the disease had been rendered quiescent, however, tsetse control measures were sporadic thereafter, and aimed at preventing re-infestation of scrub and thicket in settlement areas, chiefly by ground applications of residual dieldrin. These measures seem to have been discontinued altogether in the late seventies.

A resurgence of the disease occurred in 1980. In response, the authorities, in 1981, undertook to eradicate the fly from the whole valley and adjacent hillsides, by a programme of sequential aerial applications of endosulfan aerosol spray, supplemented by ground application of dieldrin 1.8% e.c. and limited bush clearance in areas of difficult terrain for aircraft. Despite nine aerial sprays at 12-day intervals of relatively heavy dosages of insecticide (19-38 g a.i./ha.), eradication was not achieved, but the population was reduced by over 99.9% by the end of operations in May, 1981. A model of spray effectiveness suggested, however, that this final reduction was the cumulative effect of only about a 90% kill per spray application, and that several thousand flies possibly remained after spraying. Post-spray monitoring later showed that the population in thicket fully recovered in little more than twelve months (21,22).

Escalating sleeping sickness incidence during the first half of 1983 resulted in a further tsetse eradication attempt, again by sequential aerial spraying but using a ULV natural pyrethrum formulation. Seven spray cycles were carried out at 14-day intervals from June 1983 onwards, with dosages increased progressively. Our monitoring failed to detect any impact whatsoever on the tsetse population.

Presently, yet another eradication attempt is being embarked upon: a 5-year programme of ground spraying operations, using residual insecticides. Starting in October 1984, four monthly applications were made to 10% of the principal thicket vegetation by spraying along cut paths, 100m apart. This was to be followed by a programme of localised spraying to mop-up any residual pockets of fly and peripheral infestations elsewhere. The insecticides used were dieldrin 1.8% e.c. and cypermethrin 0.3% e.c. The rationale for using two insecticides simultaneously is unclear; also it appeared that changes were made from one to the other depending upon the availability of supplies. By April 1985, after the first four applications, it was estimated that the population had been reduced by about 99.9%, but survivors were also present throughout. Spraying resumed in October, 1985 (and is continuing to date) and appears to involve little less than a complete re-spray.

From the above account, it can be seen that dieldrin has been used longest, but never in such a way that all of the tsetse population was subjected at one time. The endosulfan spray, on the other hand, while of limited duration, encompassed virtually the whole population, and, with nine sprays over 100 days, extended over at least two, possibly three, successive generations of the insect. It was therefore decided that the population would be tested for resistance to both dieldrin and endosulfan. Since no susceptibility tests were carried out previously, the alternative approach was undertaken, which was to compare the Lambwe Valley G. pallidipes strain with one that has never been sprayed. For this, an allopatric population from Nkruman, in the Rift Valley and 600 km from Lambwe, was chosen.

Materials and Methods

Samples of G. pallidipes taken in biconical traps (10) were sexed, transferred separately to standard tsetse feeding cages (about 15 per cage) and fed daily on rabbits, first in the field and then later in the laboratory after transport in 'cool'boxes' to Nairobi. Because the Lambwe population had been drastically suppressed by spraying it took several days for a sufficiently large sample for dosing purposes to be accumulated, whereas sufficient flies from Nkruman, which was also closer to Nairobi, could be collected and taken to the laboratory within 24 h. Flies were held 4-6 days before dosing (to screen out mortalities due to handling and transport, and to accustom the two strains to the same diet), during which time they were kept in an insectary maintained at 25°C and 80% RH. To standardise the material further, teneral flies (which are also virgin in G. pallidipes) were excluded from the samples initially, as later were pregnant females which, just prior to dosing, had a late second or third-instar larva in utero, judging by external abdominal appearance. Flies were dosed 24 h after their last feed. All flies were thus more than one week old, and females in early pregnancy, when dosed.

Concentrations of endosulfan (Thionex, 99% pure endosulfan, Makhteershim Beer-Sheva Chemical Works Ltd., Israel) and dieldrin (95% pure technical grade, Kenya Shell Ltd., Nairobi) were prepared by serial dilution in acetone from stock solution and kept at -4°C until used. Following trials to determine optimal dosage levels, flies were dosed with a range of four concentrations of each insecticide, from 4 to 32 ng/fly for males

19

and 8 to 64 ng/fly for females. Control flies were dosed with acetone only. The material was divided so that roughly equal numbers of flies were dosed with endosulfan and dieldrin on the same occasion, usually in batches of 15-30 flies per dose. Dosage was applied in the form of 1 µl of insecticide solution to the dorsal thoracic surface of individual flies using a microapplicator (ISCO, Model M Instrumentation Specialities Co., Lincoln, Nebraska, USA) fitted with a 0.25 ml syringe and a 27 gauge needle. Flies were hand-held while dosing, without prior anesthesia, returned to clean feeding cages and offered a blood-meal immediately afterwards and again 24 h later. Mortality counts were made 24 and 48 h after dosing. Data accumulated over approximately monthly intervals from August to November, 1985, was grouped, and the 48 h mortality data transformed to probits of mortality for analysis of the dose-response relationships.

To assess for physical differences between the strains, measurements were made of wing-vein lengths (the 'cutting-edge' of the 'hatchet'-shaped cell) in monthly samples of 50 dosed flies of each sex from each strain, which gave an index of size. The experimental procedure did not allow for routine weight measurements to be made of dosed flies, but on two occasions (in October and November) flies surplus to dosing requirements were available, and 50 of each sex and strain were weighed at the time the rest of the material was dosed.

Results

The mortality data and regression lines obtained with endosulfan and dieldrin for males and females of G. pallidipes from Lambwe and Nkruman are plotted in Fig. 1. Corresponding values of control mortality, LD_{50} (with 95% confidence limits, calculated by the method of Litchfield & Wilcoxon (16)) and LD_{95}, together with relative values and tests for significant differences, are shown in Table I. Control mortalities were below 5%, so there was no need to correct for these.

The mortality data and regression lines in Fig. 1 show that Lambwe flies were less susceptible to both endosulfan and dieldrin than those from Nkruman. These differences are examined in more detail in Table I. The LD_{50}'s for Lambwe flies were 1.21 to 1.54 times greater than those for Nkruman flies, and, on the basis of non-overlap of the 95% confidence limits, the differences were judged to be significant, with the exception of males dosed with dieldrin. As the regression lines were divergent (again excepting males dosed with dieldrin) the differences in the LD_{95}'s were even greater, about twice and high in males and females from Lambwe than in their counterparts from Nkruman with endosulfan and 1.73 times greater in Lambwe females than Nkruman females with dieldrin. Only in the case of females, however, were the slopes of the regression lines significantly different.

With regards to subsidiary findings, the general patterns were that males of both strains were more susceptible than females to both insecticides, especially at lower dosages (convergent regression lines with LD_{95} ratios closer to unity than LD_{95}'s), and that endosulfan was slightly more toxic than dieldrin, especially to females.

The extent to which the above differences in susceptibility between Lambwe and Nkruman flies and between males and females could have been on account of physical differences was examined by comparing mean wing-vein lengths and weights. The results (Fig. 2) show that Lambwe flies were consistently larger than Nkruman flies. Taking weight as the more meaningful indicator, males and females from Lambwe were, on average, 1.10 and 1.05 times heavier than their respective counterparts from Nkruman. Comparing these relative values with those obtained from LD_{50}'s, however, the latter

were considerably greater than would be expected if the differences in
weight alone were responsible for the differences in susceptibility between Lambwe and Nkruman flies. As regards the sexes, males from Lambwe
and Nkruman were, on average, 0.84 and 0.79 times as heavy as respective
females. The greater approximation of these relative values to the corresponding LD_{50}'s indicated that weight alone could have accounted for
much, but not all, of the difference between the sexes in susceptibility
to insecticides.

Discussion

Assuming that the unsprayed G. pallidipes population at Nkruman
represented the fully susceptible genotype and provided reasonably accurate measures of the base-line susceptibility levels of this species to
endosulfan and dieldrin, then the enhanced tolerances shown by the Lambwe
population, as shown by the flatter slopes of the dosage-mortality regression lines and the greater LD values, were prima facie evidence of some
development of specific resistance and indicative of the presence of het-
erozygotes for resistance to both insecticides in this population (5).
That this was so for both insecticides, despite the marked differences
in the history and mode of use, may have been due to the fact that they
are structurally similar chlorinated hydrocarbons in the hexachlorocyclo-
pentadiene group, in which it is known that selection in one cyclodiene
derivative induces resistance to all the rest (5). In the Lambwe Valley
situation, dieldrin and endosulfan may therefore have had a mutually re-
inforcing effect on the development of resistance.

The differences between the strains were small, however, and reasonably clearly manifested only in females. It would therefore be inappropriate, at this stage, to accept the evidence for developing resistance
unquestionably; either resistance is in a very early phase of development
(and hence naturally difficult to detect), or other variables were responsible for the differences in susceptibility. The obligation to compare
strains undoubtedly admitted certain inherent variables into the experimental situation. Genetic heterogeneity between allopatric populations
of G. pallidipes in Kenya is known from the evidence of enzyme polymorph-
ism, and manifestations of population diversity have been found in such
behavioural and physiological traits as feeding frequency, diel activity
patterns and reproductive performance (11). How far this heterogeneity
extends to insecticide susceptibility is unknown, but Hadaway et al. (12)
recorded almost identical susceptibility levels to both endosulfan and
dieldrin in teneral G. palpalis (Robineau-Desvoidy) originating from
Nigeria and Zaire.

The two strains of G. pallidipes tested here showed consistent size
and weight differences, but not of a magnitude which could have accounted
for the differences in susceptibility. Largely overriding the weight
differences between the strains were the variations in weight between
individual flies, which ranged up to two-fold. This was partly in consequence of the experimental procedure, which, despite the efforts to standardise the material, could not ensure that every fly took a blood meal
24 h prior to dosing, although most did. Nor was it possible to regulate
the amount of blood ingested. These factors, together with the size variations between samples collected on different occasions, probably accounted for much of the overall variability in the data, but there were no
grounds for supposing that this source of variability was any different
in one strain than the other.

Age, and pregnancy in females, were unlikely sources of variability
between the strains. Burnett (6,7) has shown that old males of G. morsi-

tans Westwood and G. swynnertoni Austen are no more tolerant than young
males of chlorinated hydrocarbon insecticides, and, in females, differences in susceptibility with respect to age to be more closely related to
reproductive state, i.e., whether virgin or pregnant. In the present
study, flies were at least a week old when dosed, and females in as uniform a state of pregnancy as was possible using wild material and conti-
ngent upon the need to dose adequately sized samples.

Variability in weight arising from feeding, and in age and pregnancy,
would have been avoided using teneral flies, but, short of establishing
colonies of both strains, this was not possible.

Few meaningful comparisons could be drawn between the susceptibility
data obtained elsewhere. Firstly, because no susceptibility tests have
hitherto been carried out on G. pallidipes (except for a study by Irving
(13) concerned only with females in late pregnancy); secondly, due to the
differences in a) the physiological state of the test insects, and b) the
experimental procedures. In general, however, the findings that males
were more susceptible than pregnant females to organochloride insecticides and that endosulfan was slightly more toxic than dieldrin were in
agreement with the findings of Burnett (7,8), Hadaway et al. (12) and Kwan
& Gatehouse (14) for other tsetse species.

If the propositions are acceptable that the isolation of the Lambwe
Valley G. pallidipes population, together with the long history in insecticide usage, are conducive to the development of resistance in tsetse
flies, and that the results of the present investigation were at best
equivocal in this respect, then, as far as conventional tsetse control
measures in general are concerned, resistance to insecticides is unlikely
to pose a serious problem, now or in the immediate future. Even if resistant individuals arise in a population subjected to spraying, the seas-
onality of most spraying operations allows the population time to revert
to the susceptible genotype, and, as most sprayed areas are subject to
re-invasion, immigrants would further dilute the resistant gene pool.
Conceivably, this situation could change as a result of recent innovations in tsetse control. Namely, the use of insecticide-impregnated traps
and screens (targets)(15,20,19). More intensive selection pressure arising from the deployment of these devices on a continuous basis and (as is
now being contemplated) over large areas, is a possibility which should
not be ignored, and suggests the need for more routine susceptibility
studies on pre- and post-control tsetse populations than has hitherto
been the case.

REFERENCES

1.   ALLSOPP, R. (1984). The control of tsetse flies (Diptera:Glossini-
     dae) using insecticides: a review and future prospects. --Bull.
     ent. Res. 74, 1-23.
2.   BALDRY, D.A.T. (1971). The control of Glossina pallidipes in East
     Africa by the aerial application of dieldrin invert emulsion spray.
     --pp. 311-325 in Int. Sci. Coun. Trypanosomiasis Res. Control.
     Thirteenth Meeting. Lagos. 1971.
3.   BALDRY, D.A.T. (1972). A history of Rhodesian sleeping sickness in
     the Lambwe Valley.--Bull. Wld Hlth Org. 47, 699-718.
4.   BROOKS, G.T., BARLOW, F., HADAWAY, A.B. and HARRIS, E.G. (1981).
     The toxicities of some analogues of dieldrin, endosulfan and isoben-
     zan to bloodsucking Diptera, especially tsetse flies. --Pestic. Sci.
     12, 475-484.
5.   BROWN, A.W.A. and PAL, R. (1971). Insecticide resistance in arthrop-
     ods. Monograph Ser. W.H.O. no. 38, 491 pp.

6. BURNETT, G.F. (1962a). The susceptibility of tsetse flies to topical applications of insecticides. III.-The effects of age and pregnancy on the susceptibility of adults of Glossina morsitans Westw. --Bull. ent. Res. 53, 337-345.

7. BURNETT, G.F. (1962b). The susceptibility of tsetse flies to topical applications of insecticides. IV.-Wild-caught adults of Glossina swynnertoni Aust. --Bull. ent. Res. 53, 347-354.

8. BURNETT, G.F. (1963). The susceptibility of tsetse flies to topical applications of insecticides. VI.-Data on more chlorinated hydrocarbons and organophosphates, and a general discussion. --Bull. ent. Res. 53, 753-761.

9. BURNETT, G.F. (1970). Resistance of tsetse to insecticides.--pp. 486-487 in Mulligan, H.W. (Ed.). The African trypanosomiases.-950 pp. London, Allen and Unwin.

10. CHALLIER, A. and LAVEISSIERE, C. (1973). Un nouveau piege pour la capture des glossines (Glossina: Diptera, Muscidae): description et essais sur le terrain. --Cah. ORSTOM, Ser. Entomol. med. Parasitol. 11, 251-262.

11. ETTEN, J. van (1981). Population diversity in the tsetse fly Glossina pallidipes Austen --Ph.D. thesis, University of Amsterdam, 1981, 110 pp.

12. HADAWAY, A.B., BARLOW, F. and TURNER, C.R. (1976). The susceptibility of different species of tsetse flies to some insecticides.--Misc. Rep. Centre Overseas Pest Res. no. 23, 14 pp.

13. IRVING, N.S. (1968). The absorption and storage of insecticide by the in-utero larva of the tsetse fly Glossina pallidipes Aust.--Bull. ent. Res. 58, 221-226.

14. KWAN, W.H. and GATEHOUSE, A.G. (1978). The effects of low doses of three insecticides on activity, feeding, mating, reproductive performance and survival in Glossina morsitans morsitans (Glossinidae). --Ent. exp. and appl. 23, 201-221.

15. LAVEISSIERE, C. and COURET, D. (1980). Traps impregnated with insecticide for the control of riverine tsetse flies.--Trans. R. Soc. top. Med. Hyg. 74, 264-265.

16. LITCHFIELD, J.T. and WILCOXON, F. (1949). A simplified method of evaluating dose-effect experiments.-- J. Pharmacol. 96, 99-113.

17. LEROUX, J.G. and PLATT, D.C. (1968). Applications of dieldrin invert emulsion by helicopter for tsetse control.--pp. 219-229 in Int. Sci. Coun. Trypanosomiasis Res. Control. Twelfth Meeting. Bangui. 1968.

18. MAUDLIN, I., GREEN, C.H. and BARLOW, F. (1981). The potential for insecticide resistance in Glossina (Diptera: Glossinidae)-an investigation by computer simulation and chemical analysis.--Bull. ent. Res. 71, 691-702.

19. OLADUNMADE, M.A., TAKKEN, W., DENGWAT, L. and NDAMS, I. (1985). Studies on insecticide-impregnated targets for the control of riverine Glossina spp. (Diptera: Glossinidae) in the sub-humid savanna zone of Nigeria.--Bull. ent. Res. 75, 275-281.

20. POLITZAR, H. and CUISANCE, D. (1984). An integrated campaign against riverine tsetse, Glossina palpalis gambiensis and Glossina tachinoides, by trapping, and the release of sterile males.-- Insect Sci. Applic. 5, 439-442.

21. TURNER, D.A. (1984). A preliminary assessment of some immediate and long-term effects of aerial spraying of endosulfan on Glossina pallidipes (Austen) in the Lambwe Valley, Kenya.--Insect. Sci. Applic. 5, 425-429.

22. TURNER, D.A. and BRIGHTWELL, R. (1986). An evaluation of a sequential aerial spraying operation against <u>Glossina pallidipes</u> Austen (Diptera: Glossinidae) in the Lambwe Valley of Kenya, aspects of post-spray recovery and evidence of natural population regulation.--Bull. ent. Res.

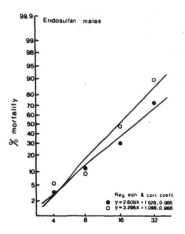

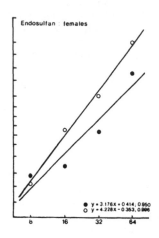

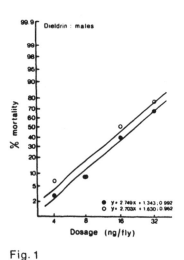

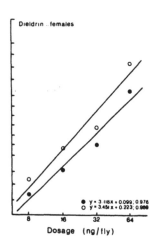

Fig. 1

Dosage-mortality data and regression lines for topical applications of endosulfan and dieldrin to wild-caught, non-teneral males and females of *G. pallidipes* from Lambwe Valley (●) and Nkruman (O). Regression equation: probit *y* = m log *x* + c.

TABLE I. *Values (ng/fly) of LD_{50} and LD_{95} for topical applications of endosulfan and dieldrin to wild-caught, non-teneral males and females of Glossina pallidipes from Lambwe Valley and Nkruman, with relative values (ratios) and tests for significant differences*

Insecticide	Sex	Source	No. tested per dose	% control mortality	LD_{50}	95% c.l. of LD_{50}	LD_{95}	Ratios Lambwe : Nkruman LD_{50}	LD_{95}	$F^{†}$	males : females LD_{50}	LD_{95}	F	Endosulfan : Dieldrin LD_{50}	LD_{95}	F
Endosulfan	males	Lambwe	105	2.20	21.41	18.10—25.32	91.43	1.40b	1.90	3.87	0.77	1.00	4.58	1.00	1.08	0.83
		Nkruman	110	2.73	15.29	13.43—17.41	48.24				0.83	1.07	12.94*	0.87	0.67	2.85
	females	Lambwe	105	2.86	27.79	24.21—31.90	91.57	1.51b	2.03	26.63**				0.74b	0.73	0.75
		Nkruman	105	0.95	18.45	16.64—20.46	45.18							0.76b	0.62	26.82**
Dieldrin	males	Lambwe	105	1.90	21.38	18.23—25.07	84.79	1.21	1.18	0.23	0.57b	0.67	4.69			
		Nkruman	105	3.20	17.66	15.47—20.16	71.70				0.73b	0.99	5.54			
	females	Lambwe	105	1.90	37.31	32.42—42.93	125.70	1.54b	1.73	8.86*						
		Nkruman	105	0.95	24.23	21.34—27.51	72.61									

b Significant difference (overlapping 95% confidence limits).

† $F_{1,4}$ Variance ratio, test for significant difference between regression coefficients (Fig. 1).

*P<0.05; **P<0.01

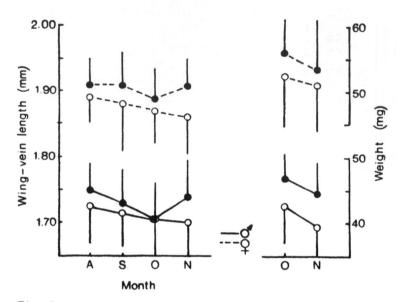

Fig. 2

Monthly mean wing-vein lengths and weights of males and females of G. pallidipes from Lambwe Valley (●) and Nkruman (o)

Tsetse control by chemical means in Malawi, Mozambique, Zambia and Zimbabwe

D.F.Lovemore
Regional Tsetse and Trypanosomiasis Control Programme of Malawi, Mozambique, Zambia and Zimbabwe

Summary

Chemical methods have been used extensively to control tsetse flies in
the Malawi, Mozambique, Zambia and Zimbabwe region for many years, especially ground spraying and to a lesser extent in the last two mentioned countries, aerial spraying. Considerable successes have been
achieved despite the reinvasion problem. More recently interest has
centred on aerial spraying carried out at night when reinvasion conditions are optimal using endosulfan at low dosages. Results have been
so encouraging that serious consideration is being given now to progressive operations using this technique in conjunction with odour baited and insecticide treated targets to eradicate tsetse flies from
the entire area of the common fly-belt. The extent of this area is
approximately 322 000km^2. The purpose of the targets would be to limit
reinvasion and provide support in the more difficult terrain. The feasibility of such an ambitious undertaking is currently being investigated under the European Development Fund financed Regional Tsetse
and Trypanosomiasis Control Programme of Malawi, Mozambique, Zambia
and Zimbabwe.

1. Introduction

The main chemical methods used in the region to date to control the
two economically important species, <u>Glossina</u> <u>morsitans</u> Westw. and <u>G.
pallidipes</u> Aust. have been the two ground spraying techniques, knapsack and
Unimog spraying and aerial spraying in its various forms, with the major
effort having been in Zambia and Zimbabwe. There has been limited usage of
chemicals in the other two countries only, namely a small knapsack spraying
trial against <u>G. brevipalpis</u> Newst. in Malawi in 1959, Nyironga (1), using
dieldrin and control operations against <u>G. morsitans</u> and <u>G. pallidipes</u> in
the Muabsa and Ria Save areas of Mozambique during the period 1962-1975
using dieldrin until 1969 and then DDT thereafter.

Ground spraying is essentially the discriminative application of persistent insecticide from the ground. The insecticide is applied to the
resting and refuge sites of tsetse flies within those vegetation elements
of the general habitat known to be favoured by the fly during the very hot
weather period which normally preceeds the end of the dry season. Generally
the vegetation favoured is that associated with surface drainage systems,
that occurring around geological features and that found along contacts
between different soil types. In Zimbabwe this favoured vegetation is referred to as essential habitat. The resting and refuge sites are treated
with insecticide during the cooler months of the dry season so that a lethal deposit is in place by the time the tsetse fly are compelled to retire
to them. In general, the essential habitat does not exceed ten per cent of

27

the total habitat. Occasionally, however, situations do occur where the vegetation is so uniform in appearance that it is not possible to differentiate which particular elements are important to the tsetse's survival, thus necessitating more extensive spraying. It has also been found useful to treat resting sites located along road and track verges, well-used game and cattle paths, field edges and around established cattle pens, dips and inspection races.

Aerial spraying, on the other hand, involves sequential ultra-low-volume (ULV) blanket applications of a non-persistent insecticide to the habitat as a whole from low-flying aircraft openating either in daylight hours or at night. Intervals between applications and the number needed to achieve eradication are related to the temperature-dependent periods required for newly emerged females to mature their first larvae and the length of the pupal period, respectively.

Although both forms of ground spraying are efficient methods of control, considerable concern is currently being expressed about the continued use of persistent insecticides. Other than dieldrin and DDT, however, there are no suitable alternatives. It would seem though that aerial spraying with endosulfan presents less of a threat to the environment.

2. The chemical techniques

2.1 Knapsack spraying
The technique was developed in West and East Africa during the nineteen-fifties.

Other than a small trial in 1958, it was first used successfully in Zimbabwe in 1960 in an area of $258km^2$ situated near the western end of the Zambezi fly-belt. The operation came to be known as the Maseme Experiment. Since then the method has been developed to its present level of sophistication and capacity. Approximately 50 000km^2 of infested area were reclaimed within Zimbabwe during the period 1961-1975, including the area cleared by joint operations conducted in the Mozambique-Zimbabwe border region between the Sabi (Rio Save) and Limpopo rivers by the governments of Mozambique , the Republic of South Africa and Rhodesia. Further successful large-scale operations have been conducted since Independence in 1980 resulting in the clearance of approximately 10 000km^2 of tsetse infested country (Hursey, pers.comm.).

On the other hand, the method has only been used to a limited extent in Zambia where there is a preference for Unimog spraying. The Department of Veterinary and Tsetse Control Services possesses a small knapsack spraying capacity, but this has not been utilised in recent years.

Two forms of knapsack spraying are employed in Zimbabwe. The most commonly used is known as conventional spraying and the other is parallel-line spraying. Conventional spraying is used wherever the essential habitat is readily distinguishable on both aerial photographs and the ground and parallel-line spraying in those few situations where the essential habitat cannot easily be determined in view of the nature of the vegetative cover. Spraying in this case is based on a grid of parallel lines spaced 200m apart.

A prerequisite for successful knapsack spraying is adequate access. In Zimbabwe, it is considered that a spraying team should never work further than three kilometres from the vehicle carrying its insecticide stock. Access should therefore be planned and developed well in advance of an operation. Planning is effected on aerial photographs in order to ensure a balanced coverage of the area to be sprayed.

A second prerequisite is an exact spraying plan. This is again prepared from aerial photographs. The planner marks all the essential habitat

visible on the photographs with a wax pencil, together with roads and
tracks, well-used game and cattle paths, field edges, established cattle
pens, dips and inspection races using a stereo pair. He also indicates the
areas requiring parallel-line spraying, if any. The marked photographs are
subsequently issued to the leaders of the spraying teams, who are required
to follow these closely.

Planning of spraying operations in Zambia is carried out on 1:50 000
scale maps. Despite their excellence, they do not provide the detail necessary when seeking out the essential habitat in the spraying areas.

The insecticide used in all knapsack spraying operations in Zimbabwe
up to and including the 1967 season was a 3,1 per cent dieldrin emulsion.
In the light of West and East African work, a start was made in 1968 to
convert to the very much cheaper DDT. This was applied as a suspension containing 5 per cent active ingredient (ai). The change was completed during
the 1969 spraying season and DDT has been used exclusively since. More
recently the concentration of ai was reduced to 4 per cent, Hursey(2).

The decision to switch was only made after the longevity of the residual properties of DDT had been clearly demonstrated locally. In trials
conducted under the extreme conditions of the Zambezi Valley at Rekomitjie
Research Station, this was found to be well in excess of eight months. DDT,
in fact, proved to be superior to the dieldrin being tested simultaneously,
Vale(3). The financial saving which resulted enabled an almost doubling of
the Branch of Tsetse and Trypanosomiasis Control's spraying capacity. Consequently, it became possible to operate over more extensive areas than
had hitherto then been possible.

In the case of Zambia, MacLennan(4) recommended in 1975 that the concentration be reduced from 5 to 3.75 per cent, but this had not been followed.

As regards the staffing requirements, equipment used and spraying procedure in the field these have been adequately described by Robertson and
Kluge(5) and it is not proposed to repeat these details here. It is suffice
to say that there has been little change in these over the years since the
Maseme Experiment, other than converting from motorised to pneumatic sprayers.

Insecticide application rates for spraying carried out in Zimbabwe and
during the joint operations conducted in the Mozambique-Zimbabwe border
region have been given in Table I. It will be noted that the amount of ai
applied for conventional spraying varied considerably. The initial increase
was attributed to denser essential habitat and broken terrain. It was also
suggested that the developing shortage of experienced planning and super-
visory staff was contributing to the problem. Staff were urged to try to
reverse the upward trend in the 1975 operations, which led to an improvement (207 to 191 g/ha). Unfortunately, the level rose again in 1976, 1980
and 1982 (and 1983) to as high as 255 g/ha, but in 1984 and 1985 it fell
below the 200g level following the reduction in concentration to 4 per cent
ai.

With parallel-line spraying application, rates are normally considerably higher, the level depending on the density of the habitat treated.

MacLennan reported in 1975 (4) that the current application rate of
insecticide for knapsack spraying in Zambia was 640 g/ha.

The knapsack spraying technique is suitable for most forms of habitat,
excepting that which occurs in broken to very rough terrain. In such situations, the development of access is costly and frequently difficult to
construct, to the extent that it is not always possible to obtain the desired coverage. As a result insecticide carriers have not only steep slopes
to negotiate in these areas, but they also frequently have to walk much
greater distances. The variable and often disappointing spraying results in

Table I. Knapsack spraying data for Zimbabwe and the joint operations conducted in the Mozambique-Zimbabwe border region between the Sabi (Rio Save) and Limpopo rivers during the period 1971-1974, inclusive and for Zimbabwe alone for the years 1975 and 1976 and 1980 and 1982. The insecticide dispensed was DDT in the form of a suspension containing 5 per cent active ingredient (ai)

Year	Total area treated km²	Total volume of DDT suspension applied to treated area litres	Average rate of application of DDT suspension/ litres/km²	Total mass of 75% WP applied to treated area kg	Total mass of ai applied to treated area kg	Average rate of application of ai/km² kg/km²	Average rate of application of ai/ha g/ha
1. Conventional spraying							
1971	7 897	2 187 469	277	145 831	109 373	13,85	139
1972	10 877	3 535 025	325	235 668	176 751	16,25	163
1973	10 585	3 784 955	358	252 330	189 248	17,88	179
1974	8 167	3 380 703	414	225 380	169 035	20,70	207
1975	9 150	3 497 625	382	233 175	174 881	19,11	191
1976	8 454	3 693 894	437	246 260	184 695	21,85	219
1980	5 246	2 655 144	506	177 010	132 758	25,31	253
1982	8 313	4 236 595	510	282 440	211 830	25,48	255
2. Parallel-line spraying							
1971	343	289 835	845	19 322	14 492	42,25	423
1972	373	207 388	556	13 826	10 370	27,80	278
1973	384	203 625	530	13 575	10 181	26,51	265
1974	392	253 251	646	16 883	12 662	32,30	323
1975	No parallel-line spraying carried out						
1976	342	85 158	249	5 677	4 258	12,45	124
1980	179	124 231	694	8 282	6 212	34,70	347
1982	11	7 986	726	532	399	36,27	363

Notes: 1. The years 1977-1979 inclusive have been omitted as operations were frequently disrupted by security requirements.
2. The year 1981 has also been omitted because the concentration of the insecticide was varied in some areas, as previously mentioned, making comparisons difficult.
3. Figures have been rounded where appropriate.

the Zambezi escarpment area can be attributed to a large extent to the extremely difficult working conditions which exist there.

The cost for Zimbabwe's 1982 knapsack spraying operations was $134/km^2, inclusive of salaries, wages and subsistence allowances for all grades of staff (42%), transport (16%), insecticide (40%) and miscellaneous items (2%). Costs of access were not included. The 1985 cost was $185,55. The increase is related to the devaluation of the Zimbabwe dollar.

The most recent comparable cost for Zambia's operation was K452/km^2 for the 1980 operations, being wages and subsistence (53%), insecticide (30%), and transport and miscellaneous items (17%). Once again costs of access development were not included. Such high levels of expenditure are regarded by officials of the Department of Veterinary and Tsetse Control Services as a major obstacle to extending the method. Nor were they able to visualise any way of effecting economies, particularly in the light of the high prevailing labour costs.

A major concern regarding knapsack spraying is the pollution caused by the two persistent insecticides used, dieldrin and DDT. Dieldrin is extremely toxic to fish, reptiles and amphibia, even in the form of a few small droplets from the finest spray emitted from the nozzle of the sprayer which might fall on the surface of a shallow pool of water containing such creatures. Both insecticides have the insidious property of bio-accumulation, where the chemical is progressively accumulated in different steps of food chains. There was a feeling of relief when dieldrin was replaced by DDT in Zimbabwe in 1968, but this was shortlived as World concern about DDT grew. It was generally argued though by glossinologists that the level of application of DDT for tsetse control purposes was comparatively insignificant in relation to the quantities dispensed for agricultural purposes. These are often thirty times as much in any single growing séason, according to Cockbill(6). It was also repeated on an annual basis as compared to the usually single application required for tsetse control operations which normally progress in annual phases.

More recently, however, Thomson (7) reported serious environmental contamination in Zimbabwe by DDT, which he related to agricultural and tsetse control areas. His findings were based on the level of residues of DDT and its derivatives in the eggs of various raptors, notably the fish eagle (Haliaetus vocifer). He also expressed anxiety for the fish populations of Lake Kariba, which body of water is fed from areas in Zambia and Zimbabwe where DDT has been used for both agricultural and tsetse control purposes. Considerable pressure was brought to bear on the Zimbabwe Government to terminate the use of DDT, except possibly for malaria control, as a result of this report.

Instead, it was decided to deregister the chemical for agricultural purposes, but to continue its use for malaria and tsetse control until satisfactory alternatives could be developed. In the case of the latter a "phasing out within five years" was mooted but has never been confirmed. At the same time, the monitoring programme of DDT residues and those of its derivatives was to be intensified. This led to the Branch of Tsetse and Trypanosomiasis Control, which was a member of the Pesticides Committee of the Agricultural Research Council of Zimbabwe and the Natural Resources Board approaching the Centre for Overseas Pest Research (now Tropical Development and Research Institute (TDRI)), Overseas Development Administration, United Kingdom, for assistance. The request was met in the form of an experienced Scientific Officer, Dr. P. Matthiessen. Dr. Matthiessen conducted a preliminary survey in which he concluded that whilst the levels of DDT residues in the study area were not sufficiently high to warrant the discontinuation of DDT on conservation or human safety grounds they did pose a potentially serious problem which justified further investigation of biological impacts (8). Plans are in hand to continue this work.

Similar concern exists in Zambia about the continued use of DDT.

Another concern of ecologists about knapsack spraying in Zimbabwe is its possible effects on vegetation succession. It is routine practice to burn the grass ahead of spraying teams, other than on wildlife land and commercial farms which might fall within the spraying area. It is seldom possible to respray an area which has been burnt towards the end of the dry season, because the spraying teams have usually been disbanded at that late stage.

Although knapsack spraying has proved a very effective means of eradicating G. morsitans and G. pallidipes, its future use is uncertain. This is partly due to the considerable organisation and effort required to achieve the successful placing of the persistent insecticide in position before the onset of the extreme hot period of the dry season and partly to the damage which it could be causing to the environment.

2.2 Unimog Spraying

Zambia has achieved considerable mechanisation in ground spraying by utilising a Mercedes Benz Unimog tractor model 411 fitted with spraying equipment and supported by a second Unimog carrying water.

Planning for Unimog spraying is effected on 1:50 000 scale maps. These are then issued to staff supervising the operations in the field. Marking of the routes which the Unimog will follow, called spray-lanes, is carried out in advance of the operations. These spray-lanes follow the essential habitat and other situations which are planned for treatment. Marking is done by blazing trees.

The insecticide used in the first instance in an operation is DDT in the form of a suspension containing 5 per cent ai. Towards the end of the dry season the unit converts to a mixture of 3,2 per cent DDT ai suspension and 1,8 per cent dieldrin emulsion. It is believed that dieldrin deposits are more resistant to the effects of rain than DDT. This innovation is interesting, but the need for the conversion questionable. Once the first rains have fallen and the vegetation is flushed, tsetse flies are no longer dependent on the essential habitat for survival. They can then wander at will throughout the general habitat. The position has been discussed with Evison, who said their reason for adopting this system was that they invariably only commenced spraying operations well into the dry season. This necessitated teams having to continue working through to the rains in order to complete the areas allocated for treatment and allowing little time for the insecticide to be effective before rains washed it away.

Zambia has four Unimog spraying units, each comprising a vehicle fitted for spraying and another for carrying water. The annual capability of each unit is approximately 250 square kilometres, based on a five month working period.

Spraying equipment includes a 750 litre insecticide tank, a powerful Platz spray pump and two variable spray lances. The pump and a stirring device within the insecticide tank are operated from the power take-off of the vehicle. Swivelling tractor-type seats mounted at the rear of the vehicle are provided for the two operators.

Spraying procedures involve the Unimog moving slowly along the spray-lane guided by two scouts pointing out sites which require treatment to the two operators. The insecticide is applied under considerable pressure, either as a jet or a fine spray depending on the distance between the lance nozzle and the object being treated. Although the operators are accurate

32

with their applications, it is inevitable that a considerable proportion of insecticide misses the target, particularly when the spray is used.

The application rate of insecticide was 734 g/ha when DDT was used alone and 639g DDT and 159g dieldrin/ha when the "mixture" was being used at the time of MacLennan's 1975 study (4). He advocated a reduction to as near 200 g/ha as possible. Subsequent rates have been approximately 650 g/ha when DDT was being used alone and 566g DDT and 138g dieldrin/ha when the "mixture" was being used.

The most recent cost available was that for 1980 of $K481/km^2$. This included all salaries, wages and allowances (30%), insecticide in the form of DDT and dieldrin (43%), and running costs and other items (27%). The use of dieldrin increased the cost by about six per cent.

The technique is suited to gently undulating drainage line interspersed woodlands which predominate on the Zambian plateau and to the flatter areas of the Zambezi Valley. It is not a satisfactory method for rough terrain. It also suffers from the same problems and drawbacks of knapsack spraying which together with the current cost of new Unimogs could limit its future use.

No Unimog spraying operations have been carried out since 1982 for financial reasons.

2.3 Aerial Spraying

Aircraft were first employed as a means of applying insecticide for the control of tsetse flies as early as December, 1945, by South Africa's Division of Veterinary Services in Zululand, now KwaZulu. Initially the insecticide was applied through a venturi apparatus mounted beneath the fuselage. This technique was soon changed for one in which the insecticide was fed into the exhaust stack of the aircraft. This emitted the spray as a dense white aerosol of coarse droplets. The insecticide used was 16 per cent DDT ai dissolved in diesoline, later replaced for reasons of economy by BHC. The formulation used was 4 per cent BHC (w/v) in diesoline. Applications were most satisfactory under conditions of a strong temperature inversion with minimal wind and turbulence. Other than during the night these occurred for only about two hours after dawn and one hour before sunset. As a result, operations were restricted to the early morning and evening, which seriously limited the full utilisation of the aircraft.

Aircraft guidance was based on the visibility of the aerosol applications. The leading aircraft laid the first swath with the second and subsequent aircraft taking up positions in echelon formation to windward. This procedure was repeated until the particular area to be treated was completed, or until the atmospheric conditions were no longer suitable. A variety of aircraft were employed including J.U.52s, then Avro Ansons and finally Piper Cruisers. These operations were successful, Du Toit(9).

The Colonial Insecticide Research Unit, later re-designated the Tropical Pesticides Research Institute (TPRI), operating initially over islands in Lake Victoria from Entebbe, Uganda, and subsequently at both Arusha and Kikore, near Babati, Tanzania, worked continuously from 1949 until 1964 on the development of methods of aerial application of insecticides for tsetse control purposes. These included application of dusts, testing of systems evolved by the South Africans and variations thereon, culminating in the development of the successful ultra-low-volume (ULV) technique. The atomising equipment utilised the aircraft's engine exhaust and was described by Lee and Miller(10) as a thermal exhaust aerosol generator. It was employed successfully against G.morsitans, G.swynnertoni and G.pallidipes, Hocking et al, (11), using only four applications of endosulfan at 21-day intervals at rates of 13,5 litres/km^2 of a 20 per cent formulation. The endosulfan was dissolved in an involatile oil. The diameter of the majority of the droplets was in the 20-50 microns range. Burnett(12) noted that the insecticides used in this technique must be highly toxic because of the small size of the droplets. At that time, only dieldrin, izobenzan and endosulfan met this requirement, the last being the most suitable. Applications were restricted to the early morning and evening periods. Because of the invisible nature of the aerosol, aircraft guidance had to be effected from the ground.

The first time aircraft were used to apply insecticide for tsetse control purposes in the region was in Zimbabwe in late 1953, early 1954. The area treated lay west of Karoi and was 50km^2 in extent. The contractors responsible for the latter half of the Zululand spraying undertook this operation using the same technique and equipment as previously. The operation was a failure due to unsatisfactory weather conditions.

It was repeated during the dry season of 1954, but over a more extended area, 1 000km^2. It again proved unsuccessful due to the inability of the five aircraft to complete the area designated for treatment from the third cycle due to poor weather conditions. In the final event, only about 50 km^2 of tsetse habitat received the requisite number of applications.

Two similar operations were conducted in Zimbabwe, the first at Kariba in 1956 and the second in the Lubu valley (middle Sebungwe river) in 1957, using a local contractor. Three De Havilland Tiger Moths were employed in the first operation and an Avro Anson XIX aircraft and four Tiger Moths in the second. The atomising system and insecticide used were similar to those of the earlier operations. Both achieved their objective of reducing the tsetse population to a low level. Costs of the second operation, which was carefully monitored (inclusive of insecticides, hire of aircraft and personnel and supervision and general labour), were \$360/km^2 (£465 15s/sq. mile), Cockbill et al, (13).

In 1968, Zambia put the TPRI ultra-low-volume application technique to its first large-scale test over an area of 1 600km^2 in the Westen Province. The area was sprayed five times, at approximately three-week intervals, with 20 per cent endosulfan in Shellsol AB oil at an average rate of 14,4 litres/km^2/application. The application rate of the active ingredient for the operation as a whole was 15 kg/km^2 (150 g/ha). Applications were restricted to the early morning and evening with aircraft guidance being effected from the ground. Three Cessna 180 and two Piper Pawnee aircraft were employed. Eradication was achieved, except in a small area adjacent to an unsatisfactory isolation barrier, Park et al, (14).

Following the success of the 1968 operations, Zambia continued to exploit the technique for its major reclamation activities. After an unfortunate attempt to engage a spraying contractor from overseas to execute the 1969 programme, the Government decided to establish its own aerial spraying capacity. A parastatal body, Rural Air Services, was accordingly set up under the aegis of the Rural Development Corporation. Its terms of reference included charter work apart from tsetse control and crop spraying operations.

Three Beechcraft Baron aircraft were purchased for the purpose. Unfortunately these plans received a major setback when the Department of Civil Aviation declined to issue a licence for charter work, thus limiting effective aircraft utilisation to the dry season and then for only a few hours a day.

The Barons were each equipped with an insecticide tank mounted under the fuselage in line with the undercarriage and atomising equipment similar to that designed by TPRI and used in the Western Province operation in

1968. This was subsequently replaced by the more efficient Micronair rotary atomiser. Two of the Barons were fitted with Decca Doppler navigational
equipment in order to reduce dependence on ground control parties for accurate guidance. According to MacLennan(4), the three aircraft together had
an annual capability of 9 800km^2 (3 800 square miles) based on early morning and evening flying. Based on the actual performance of the Piper Aztec
of 25km^2/hour, Hursey and Allsopp(15), the three hours of daylight which
are usually suitable for spraying and a maximum of fourteen operational
days in which the aircraft is covering "new ground" each day in every three week period, it was unlikely that the real capability could have been
greater than 1 050km^2 for each aircraft, especially bearing in mind that
the flight paths were only spaced 115m apart.

It is noteworthy that the three aircraft together sprayed 1 500km^2
of tsetse habitat annually in 1970, 1971 and 1972, Kendrick and Alsop(16).
The insecticide used in these operations was 20 per cent endosulfan applied
at the rate of 10 litres/km^2/application. Eradication was achieved.

MacLennan(4) expressed considerable concern about unsatisfactory services provided by Rural Air Services, particularly in regard to skill and
reliability. He suggested the Department be freed of its obligation to
that organisation and be permitted to employ commercial aerial spraying
operators instead.

Whilst the TPRI's ULV application technique and the introduction of
electronic navigational equipment into spraying aircraft had been significant events in tsetse control, it was the advent of night aerial spraying
which really made it possible to undertake large-scale reclamation by this
means economically. The originators of the technique were Agricair (Pvt)
Ltd., a Zimbabwe-based crop-spraying organisation working in conjunction
with Hunting Surveys Botswana (Pty) Ltd. and Botswana's Department of
Veterinary Services and Tsetse Control in the Okavango Delta region in
1972. Night spraying which calls for great skill and daring, increases the
utilisation of spraying aircraft three-fold with a concomitant saving in
costs. It also creates the latitude necessary in aerial spraying to acco-
mmodate the numerous problems which occur in any operation. In the past
these had often disrupted the timing of applications, thus necessitating
additional treatments.

Zimbabwe first began investigating night aerial spraying in 1974 with
a field trial covering 210km^2. A second larger trial covering 482km^2 was
carried out in 1975, followed by a limited trial in 1981. Subsequently a
large-scale control operation was undertaken in 1982, covering 2 400 km^2,
which has been described in detail by Hursey and Allsopp(15).

The area chosen for the 1982 operation purposely included broken
country in order to test the effectiveness of the method in such situations. There had been a tendency to regard aerial spraying as being limited
to flat to gently undulating terrain previously.

Three aircraft were used, two Piper Aztecs and an Ayres Turbo Thrush.
All were fitted with powerful lights and Micronair rotary atomisers mounted below the trailing edge of the starboard wings. In addition, the Aztecs
were fitted with two types of electronic navigational equipment, Decca
Doppler and Del Norte/Decca Flagman and the Turbo Thrush solely with the
latter. The Flagman's remote transponders, which were required to be placed on the highest ground of the area being sprayed, had to be moved frequently. This was effected by helicopter.

The insecticide used was endosulfan as an emulsifiable concentrate in
Shellsol AB.

Staffing included pilots and supporting teams of the contracting
company, a senior officer of the Branch of Tsetse and Trypanosomiasis Con-

trol and the relevant staff for ground guidance teams, airstrip duties and the monitoring of tsetse populations.

The operation was undertaken during the period 27th July to 27th September, 1982. Five applications were effected, the spacing of which was closely correlated with pupal and first larval periods. An accurate check had therefore to be maintained on daily temperatures. Biological assessments based on dissection of tsetse reproductive systems were also employed as an additional check on larval development. As spraying is dependent on suitable atmospheric conditions, close attention was paid to ground and air temperatures and prevailing wind speeds. A flight path spacing of 200m was adopted with runs being aligned mostly in a north-south direction across the prevailing wind which was generally easterly, to enhance insecticide drift. Spraying progressed from east to west. While most of the spraying was at night, early morning and evening periods were utilised for treating gullies and the rougher ground, using the Turbo Thrush. Three ground aircraft guidance teams, equipped with 12m long telescopic masts carrying lighted beacons and flares, working off three east-west marker lines, provided additional guidance to the aircraft. This proved necessary as the electronic navigational equipment did not always function satisfactorily, due to the nature of the terrain. High points in the area were also marked with lighted beacons. Initially, the 2 400km^2 block took eight nights (plus daylight hours, where necessary) to complete. This was reduced to five nights (plus daylight hours) in the fourth and fifth applications, by operating the two Aztecs together, followed by the Turbo Thrush when the Aztecs were refuelling.

Application rates were: first cycle 25g ai/ha, second cycle main block 20g ai/ha, Busi block 14g ai/ha and Chizarira block 17g ai/ha, third cycle overall average 16,2g ai/ha, fourth cycle 15,4g ai/ha and fifth cycle 14,7g ai/ha. (The planned application rates had been 20g ai/ha for the first two cycles and 14g ai/ha thereafter). Total insecticide applied during the five cycles was eleven per cent higher than intended.

The operation cost $223/km^2, inclusive of flying costs (58%), insecticide (39%) and incidentals (3%). Cost of operating the airstrip, camp labour and transport were included in the incidentals. Salaries, wages and allowances of senior staff and members of the tsetse monitoring teams were not costed. Similar operations have been conducted annually since, details of which have been 1983, 2 100km^2, $275/km^2, 58,1%, 35% and 6,9%; 1984, 1700 km^2, $357/km^2, 58,8%, 36,7% and 4,5%; and 1985, 1680 km^2, $410/km^2, 53,5%, 38,5% and 8% respectively. Results have been generally satisfactory. In particular, those for the 1985 operation are very encouraging. The rapid increase in costs over the four years relates to the greater extent to the devaluation of the Zimbabwe dollar.

Zambia has unfortunately not been able to adopt a similar technique because of a ban on night flying. The matter was discussed with the Deputy Director of Civil Aviation and his Chief Operations Officer who indicated that they would probably view a request sympathetically providing the intended operation did not fall in any security sensitive area.

Aerial spraying with endosulfan at ULV application rates seems to present little threat to the environment. Magadza(17) studied the effects on non-target animal species in Zambia's 1968 Western Province operation, but found little cause for concern. Similar observations were made in Zimbabwe's 1974 and 1975 trials according to Cockbill(18). He said the numbers and varieties of arthropods, mainly insects, were studied by catches in light-traps and open trays. Mortalities amongst exposed aquatic insects, fish, frogs, mice and stingless bees (Trigona spp.) were noted. No deleterious effect of the insecticide was demonstrated, other than on fish (Tilapia spp.) in shallow water.

More recently an Overseas Development Administration team from the United Kingdom, working in the Okavango Delta region of Botswana, demonstrated the relative innocuity of endosulfan, Douthwaite et al, (19).

Night aerial spraying is currently the most promising tsetse control method for large-scale operations against G.morsitans and G.pallidipes. It does, however, require a great deal more development work, particularly relating to electronic guidance systems for aircraft and treatment of broken country.

It is essential that electronic guidance systems be perfected, as ground guidance requires the constant presence of senior staff, transport and adequate area access. The work is tedious and tiring when aircraft are operating throughout the night, week after week. A particularly attractive feature of the Flagman system is the print-out facility. This records location signals which can be transcribed to illustrate the precise course flown by the aircraft. Control of the pilot could be achieved in this manner.

It is also essential that aerial spraying be developed for use in all forms of terrain if it is to be employed to full advantages. As already indicated, ground spraying does not provide the answer in rough terrain. Immediate thought should also be given to exploiting the versatility of helicopters. Although expensive, they could be used in cases requiring special attention. Effective treatment in such situations, however costly, could mean the difference between success and failure of an entire spraying operation.

There is also the possibility that the observations made by Hursey and Allsopp(15) during the recent Zimbabwe operation might provide an answer. It seems that a great deal more aerosol reaches the ground than had been previously suspected from aircraft flying at higher levels at night than the normal spraying height. Insecticide droplet recovery and tsetse survey results from some of the more broken areas indicate that it might possibly be sufficient to achieve eradication. If this is the case, it might be possible to negotiate rough country in night spraying operations without serious loss of efficiency, simply by flying higher. Although it would require increased quantities of insecticide, this would be a small price to pay for the considerable advantage gained.

There is also the need for aerial spraying contractors to be assured of longer term employment than the single one-off operations which have generally been the case to date, if techniques are to be improved. The acquisition of suitable aircraft and electronic navigational equipment involves a considerable outlay, which the contractor must obviously recoup. Contracts of three to five years will therefore be necessary to encourage specialisation.

3. Chemical techniques and the Regional Tsetse and Trypanosomiasis Control Programme

It was the promising developments in aerial spraying which influenced the consultants responsible for the report "A Regional Tsetse and Trypanosomiasis Study:Malawi, Mozambique, Zambia and Zimbabwe", PTA Consulting Services (Pvt) Ltd., Harare, Zimbabwe (20) to recommend that serious consideration be given to large-scale progressive eradication of the 322 000 km^2 fly-belt common to those four countries, which they called the common fly-belt (Map I). They saw this as the only permanent solution to the constraint presented by tsetse flies and the disease they transmit to domestic animals and to a lesser extent man, to the full development of the region.

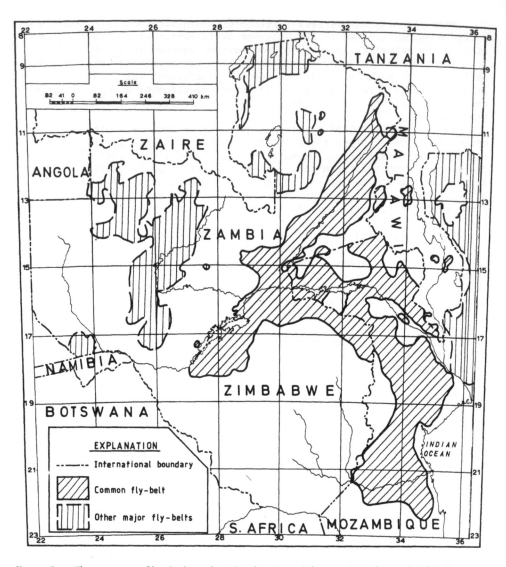

Map. I. The common fly-belt of Malawi, Mozambique, Zambia and Zimbabwe.

They appreciated the current limitations of the technique, but believed these could probably be overcome by continuation and expansion of the
research and development effort already initiated jointly by the Branch of
Tsetse and Trypanosomiasis Control of the Department of Veterinary Services, Government of Zimbabwe and the Tropical Development and Research Institute, Overseas Development Administration, United Kingdom. This work
was already beginning to show results, but required very much greater
financial and technical input than was possible at the time.

They also appreciated the significance of the host odour attractants
and target development studies being carried out jointly by the Zimbabwe
Branch of Tsetse and Trypanosomiasis Control, the Tropical Development and
Research Institute and the Tsetse Research Laboratory, Bristol, which was
then already leading rapidly to the creation of the odour baited, insecti-

cide treated, target, in the context of eradication. They saw in it a
possible means of setting up an invasion barrier, using, say a single line
of baits spaced 100m apart to use their exact words and thus eliminate the
reinvasion factor, which has dogged ground and aerial spraying endeavours
for so long. In calculating the aerial spraying progression required to
cover the 322 000 km^2, it had been considered necessary to include a 25
per cent overlap annually. Such a barrier would obviate this need. It was
also visualised that the odour baited, insecticide treated, target would
provide a means of dealing with those seemingly intractable tsetse
"pockets" which remain from time to time after progressive operations through broken terrain. Again, it was observed that progress was being ham-
pered by insufficient financial and technical input.

They also indicated their concern regarding the possible impact of
chemical techniques on the environment. They said that irrespective of the
earlier observations, the effect of endosulfan on non-target species would
need to be monitored, particularly where operations were to be carried out
in the vicinity of large bodies of water such as various man-made and natural lakes of the region or major rivers, which are the reservoirs of the
fish population. While no deleterious effects were anticipated, it was important that all possible precautions be taken. They noted that the Tropical Development and Research Institute was already involved in investigating the levels of DDT residues within the Lake Kariba drainage of Zimbabwe in relation to tsetse control operations at the instigation of the
Branch of Tsetse and Trypanosomiasis Control and the Natural Resources
Board, Government of Zimbabwe, referred to previously.

Consequently, when preparing the draft regional tsetse and trypanosomiasis control programme they included three projects relating to these
matters, namely the aerial spraying research and development, the identification of host odour attractants for tsetse flies and the environmental
monitoring projects. They also recommended that the organisations already
involved in the work should be invited to undertake these projects on an
ongoing basis of their existing work.

The three projects have been funded within the framework of the
European Development Fund financed three-year Regional Tsetse and Trypanosomiasis Control Programme of Malawi, Mozambique, Zambia and Zimbabwe,
which commenced on 1st December, 1985. The first two projects will be
undertaken by the Tropical Development and Research Institute in conjunction with the organisations already named and the environmental monitoring
project will be carried out by a European Community member states consort-
tium of environmental agencies led by the University of Saarbrücken, Federal Republic of Germany. Contracts for the first two projects have already
been signed and that for the environmental monitoring project is currently
being negotiated and should be signed shortly.

The stated objectives of the projects are, respectively, as follows:

- To determine the capabilities and limitations of aerial spraying, particularly in rugged terrain. Alternative methods will be devised for terrain in which normal fixed-wing spraying proves impossible. Other insecticides are to be evaluated.
- To identify new tsetse fly attractants for the relevant species in order to increase the effectiveness of odour baited traps and odour baited,
  insecticide treated, targets.
- To evaluate the environmental effects of pesticide use, particularly by
  the application of endosulfan from the air at low dosages during the
  three-year Regional Tsetse and Trypanosomiasis Control Programme and to
  provide an objective assessment of the likely impact of the integrated
  longer term control programme which has been mooted.

4. Conclusion

In the event of the integrated chemical control methodology which it is hoped to perfect during the three-year Regional Tsetse and Trypanosomiasis Control Programme proving successful, it is possible the Regional Standing Committee of Malawi, Mozambique, Zambia and Zimbabwe would seek the massive funding required to undertake the longer term control programme aimed at complete eradication of the common fly-belt. In this context, however, bearing in mind the need to ensure that all land reclaimed from tsetse flies is used wisely subsequently, it should be stressed that it does not really matter how long the actual clearance takes providing that this is done systematically and that when completed, it is total. Long term detailed planning and close co-operation between the four governments involved will therefore be essential.

Finally, there is always the possibility that the integrated chemical control methodology might not prove as successful as is currently hoped. In this case, whilst it would be disappointing, to say the least, the Programme should not then be judged a failure because a great deal of knowledge would have been gained and practical experience obtained, which could have considerable application throughout tsetse-infested Africa.

REFERENCES

1. NYIRONGO, M.S.P. (1982). Tsetse habitats of Malawi. Cyclostyled report of the Central Veterinary Laboratory, Department of Animal Health and Industry, Malawi
2. HURSEY, B.S. (1984). Annual report of the Branch of Tsetse and Trypanosomiasis Control, Department of Veterinary Services, Ministry of Agriculture, Zimbabwe for the year ended 30th September, 1984
3. VALE, G.A. (1968). Residual insecticides for use against tsetse flies in Rhodesia. International Pest Control, July/August, 1968
4. MacLENNAN, K.J.R. (1975). A review of operations against tsetse flies and trypanosomiasis in Zambia and recommendations for the future. Report to the Government of Zambia
5. ROBERTSON, A.G. and KLUGE, E.B. (1968). The use of insecticide in arresting an advance of Glossina morsitans Westwood in the south-east lowveld of Rhodesia. Proc.Trans.Rhod.Sci.Assoc., 53,17-57
6. COCKBILL, G.F. (1975). Handbook for the use of tsetse field staff. A publication of the Branch of Tsetse and Trypanosomiasis Control, Department of Veterinary Services, Rhodesia, Rhodesia, August, 1975
7. THOMSON, W.R. (1981). A report on chemical contamination of the environment in Zimbabwe. A report to the Director of National Parks and Wildlife Management, Zimbabwe
8. MATTHIESSEN, P. (1984). Environmental contamination with DDT in western Zimbabwe in relation to tsetse fly control operations. An ODA, U.K. report. TDRI Project No. K3007
9. DuTOIT, R. (1954). Trypanosomiasis in Zululand and the control of tsetse flies by chemical means. Onderstepoort Jnl. Vet. Res., 26,317-387
10. LEE, C.W. and MILLER, A.W. (1966). Trials with devices for atomising insecticides by exhaust gases from a Cessna 182 aircraft. Agriculture Aviation, 8,19-22
11. HOCKING, K.S., LEE, C.W., BEESLEY, J.S.S. and MATECHI, H.T. (1966) Aircraft applications. XVI Airspray experiment with endosulfan against Glossina morsitans, G.swynnertoni and G.pallidipes, Bull.ent.Res.,56 737
12. MULLIGAN, H.W. and POTTS, W.H. (1970). The African Trypanosomiases. George Allen and Unwin Ltd., London

13. COCKBILL, G.F., LOVEMORE, D.F. and PHELPS, R.J. (1963). The control of tsetse flies (Glossina: Diptera, Muscidae) in a heavily infested area of Southern Rhodesia by means of insecticide discharged from aircraft, followed by settlement of indigenous people. Bull.ent.Res., 54,93-106
14. PARK, P.O., GLEDHILL, J.A., ALSOP, N. and LEE, C.W. (1972). A large-scale scheme for the eradication of Glossina morsitans morsitans Westw. in the Western Province of Zambia by aerial ultra-low-volume applica-tion of endosulfan. Bull.ent.Res., 61,373-384
15. HURSEY, B.S. and ALLSOPP, R. (1983). Sequential applications of low dosage aerosols from fixed wing aircraft as a means of eradicating tse-tse flies (Glossina spp.) from rugged terrain in Zimbabwe. A publica-tion of the Branch of Tsetse and Trypanosomiasis Control, Department of Veterinary Services, Zimbabwe
16. KENDRICK, J.A. and ALSOP, N. (1974). Aerial spraying with endosulfan against Glossina morsitans morsitans in the Okavango Delta area of Botswana. PANS Vol.20, No. 4, December, 1954
17. MAGADZA, C.H.D. (1978). Field observation on the environmental effect of large-scale aerial applications of endosulfan in the eradication of Glossina morsitans centralis Westw. in the Western Province of Zambia in 1968. Rhod.J.agric.Res., 16 (1978)
18. COCKBILL, G.F. (1979). The effect of ultra-low-volume applications of endosulfan applied against Glossina (Diptera: Glossinidae) on popula-tions of non-target organisms in savanna woodland in Zimbabwe-Rhodesia Bull.ent.Res., 69,645-655
19. DOUTHWAITE, R.J., FOX, P.J., MATTHIESSEN, P. and RUSSEL-SMITH, A. (1981). The environmental impact of aerosols of endosulfan applied for tsetse fly control in the Okavango Delta, Botswana. An ODA, U.K. report
20. ANON. (1983). A regional tsetse and trypanosomiasis control study: Malawi, Mozambique, Zambia and Zimbabwe. A report by PTA Consulting Services (Pvt) Limited in Association with Minster Agriculture Limited to the Ministry of Lands, Resettlement and Rural Development, Zimbabwe.

Considerations concerning practical implications of the use of ivermectin to control tsetse flies in Africa

J.Mortelmans & S.Geerts
Institute of Tropical Medicine, Veterinary Department, Antwerp, Belgium

Summary

Human and animal tse-tse transmitted trypanosomiasis are still a major constraint to socio-economic development in a large part of Africa south of the Sahara. In many instances the classical methods to control trypanosomiasis and tsetse flies are too difficult or too expensive for application on the field. The search for new approaches is an urgent issue. Laboratory observations gave evidence that tsetse flies' life and fecundity could be heavily affected by subcutaneous or oral administration of a new antibiotic, ivermectin, to the host. The drug has a broad-spectrum antiparasitic activity. Especially slow release formulations could have considerable application in tsetse control operations.
It is suggested that it should be worthwhile to undertake field experiments in Africa to assess the feasibility of the laboratory findings under african field conditions.

Trypanosomiasis in both men and animals is an intractable disease which has been studied for over 80 years and still no overall solutions became available.

Animal trypanosomiasis transmitted by tsetse bite is a major constraint to development in an area south of the Sahara covering about 10 million square kilometres, or roughly one third of the African continent. Large areas where crops cannot be grown, are devoted to ruminant livestock which can transform herbage into high quality protein,rich foodstuff and other valuable by-products. In other areas where mixed farming systems are developed by small livestock owners, ruminants and other grasseating animals provide also draught and manure which are essential to assure sufficient valuable yield of foodcrop for autoconsumption and selling. It is generally estimated that trypanosomiasis is in many of these areas the main factor which limits high production levels in livestock, which is responsible for low fertility rates in domestic livestock and high mortality rates in young animals as well as in adults and subsequently for unrational exploitation of natural resources and land. Today Africa harbours about 160 million cattle, only 20 million of them living in tsetse infested areas. Should trypanosomiasis become under control or even be eradicated, livestock scientists estimate that a further 120 million cattle could be raised in these areas. Available information gives evidence that about 25 million doses of trypanocidal drugs are currently used every year to prevent or to cure the disease in domestic livestock in

Africa; the cost of this annual intervention amounts to approximately 40 million US dollars.

Human African trypanosomiasis still exists in many foci spread all over the tsetse belt. WHO officials estimate that about 45 million people are at risk, 20.000 new cases occuring each year. It is also a well known fact that the disease may destroy whole communities; in other cases large parts of land suitable for agriculture are abandonned by people fearing to become infected by the disease.

Careful field research learned that 38 african countries are infected by tsetse flies (10). These pest are represented by 30 species and sub-species, 7 out of them being of real high importance for transmission of the disease to human beings and livestock. Some other species only play a minor or even completely negligible role as they feed mainly or exclusively on game animals.

Human and animal trypanosomiasis are cured by chemical drugs, some times with doubtfull success; today chemoprophylaxis is largely and successfully used in large domestic animals; old drugs still seem to be effective, being used twice to six times a year; the price of such an annual prophylaxis scheme varies from 3 to 10 million US dollars, which makes it financially unrealistic to most of the african cattle owners. Trypanotolerance is claimed to be a solution for many West-African countries harbouring cattle breeds belonging to the taurin type (12). This trypanotolerant phenomenon is recently also demonstrated in east african zebu type (13). Trypanotolerance is giving hope to get a useful low-cost method of raising livestock resistant to the disease provided that some conditions are fullfilled.There is still little hope for a suitable immunization method to control trypanosomiasis, but it will not be for the near future.

Tsetse flies are fighted by different ways: bushclearing, traps, insecticides applicated on screens or sprayed by different methods (knapsack spray, helicopter or fixed wing aircraft spray, etc.).

The world became conscious of the potential danger of the use on large scale of highly toxic remanent insecticides; biologists and ecologists now claim for studies of the impact of these drugs on the ecological system and to stop to use them. Game shooting has been largely popular in some countries of Eastern and South Africa, but is fortunately abandonned now. Sterile male release is condamned to small scale applications as nature self handicaps large number production of offsprings.

In this context it becomes evident that new and alternative methods to control tsetse flies in Africa are urgently needed. Beside successful field experiments carried out with several kinds of screens and traps, impregnated with modern and less dangerous drugs such as pyrethroids, experiments have been set up with ear tags impregnated with insecticides. Attractants and hormones are used.

The introduction of a new very potent broad-spectrum antiparasitic drug, ivermectin, affecting most of the important immature and mature roundworms of the intestinal and respiratory tract of domestic livestock (6, 7) and even some filaroid worms (11) gave hope that it also could have an effect against flies when it became evident that the drug had very promising activity against blood sucking insects and acarids (1,2).

Avermectins are antibiotics, isolated from a newly described actinomycete, Streptomyces avermitilis (3). Ivermectin is a mixture of two synthetic derivates of Avermectins. In 1983 one of us participated in an experiment which could demonstrate the systemic efficacy of ivermectin in guinea pigs and goats infected with Glossina palpalis palpalis (4). One single bloodmeal with ivermectin already affected the flies' life expectancy. The effect started at a concentration between 1 and 2 mpk (milligramme pro kilo), but was absent at the usually recommended clinical dose of 0.2 mpk to treat endo-and ectoparasites of animals. The effect raised with increasing dosage and most flies died at 10 mpk. All flies before dying, showed a progressive paralysis, firstly reducing the flight movements, finally affecting the legs.

The idea to affect tsetse flies through action on the host is not new. Already in 1971 and 1973 it was found that some products in the diet of the host resulted in reduced fecundity when Glossina morsitans morsitans were fed on these animals (16, 17). Coccidiostats in the diet of rabbits affected tsetse flies when fed on them (8).

When the experiments in Antwerp were carried out treating the hosts by subcutaneous infection of ivermectin, it was shown in 1984 in Bristol that oral treatment could be applied also and that in this case lower dosage levels as in subcutaneous treatment could be effective (9). Twice the clinical dose (0.4 mpk) given orally to a horse, was adequate to reduce tsetse fecundity to zero following a single meal. Recently the Antwerp group (18) observed significant reduction of pupae, disturbance in puparium formation and a high proportion of females in a less advanced reproductive cycle when tsetse flies were fed on rabbits treated with ivermectin.

Both in Antwerp and Bristol it was demonstrated that tsetse flies fed several days after the administration of ivermectin , were still affected dose depending and that repeated feeding on the blood of a treated animal reduced considerably the dose of ivermectin required to produce a given effect.The lethal concentration was even reduced to < 0.04 µg per ml of blood for teneral males when fed repeatedly on treated pig blood. Pregnant female flies fed continuously on blood from a treated horse produced no offspring when fed on blood taken 8 days after treatment and only fully recovered when fed on blood taken more than one month after treatment (9). All these effects could be enhanced if slow release boluses or implants could be developped. By applying this type of formulations, the glossina population in a given area could decrease sensibly,since reductions on fecundity are not easily tolerated by insects with a low reproductive potential. Simultaneously important reductions in tick populations could be expected as is shown by Drummond et al. (5) and Pegram en Lemche (15). From this latter publication and also from the work of Parent and Samb (14) it appears that ivermectin may also have a growth promoting effect. If similar improvements in liveweight gain could be demonstrated with slow release formulations of ivermectin, then the economic treshold of the therapy would be enhanced.

All these findings merit to be tested under field conditions in Africa. Several factors have to be taken into account:
- the species of tsetse fly or flies in a defined area
- the density of the population and its seasonal variation.
- the host preference of the tsetse flies
- the density and species of man and domestic animals in a given area
- the density and species of game animals in a given area
- the facilities to treat domestic animals by injections or orally(individually, or by lick-stones)

- the feasibility to administer the drug to certain wild animal species by lick-stones.
- the price of the product and its administration
- the availibility of funds
- the beneficial effect of treatment on other internal and external para-sites
- the frequency of treatment
- the willingness of the animal owners to collaborate

All these prerequisites claim for an answer before a project should be launched.

Nevertheless it should be worthfull to identify a few feasi-ble and useful projects to test the effect of the administration of ivermectin in the field:
- an area with cattle and without or practically without game
- an area with cattle and big game
- an area with cattle and warthog

In all thse cases it should be necessary to carry out a care-ful survey on the tsetse fly population by means of traps and to do blood-meal analyses. It is indeed obvious that in an area where tsetse flies feed nearly exclusively on cattle the chances for success will be higher as in an area where a high percentage of the flies feed on game. In the area where game is present it should be interesting to know the interaction with lick-stone administration of the product. It should be ideal if these experi-ments could be carried out under high and low tsetse pressure. In all these cases the influence on the trypanosomiasis should be carefully monitored (presence of parasites, anemia, body weight etc.) as well as the influence on the tsetse population. A cost-benefit analysis of the experiments should be highly desirable.

Although the commercial price of the product is still to be considered as a constraint to applications in the field in Africa, it nevertheless could be very interesting from a scientific, biological and socio-economic point of view to get information on the practical feasibi-lity of such a program under african field conditions. In the case of a positive output, it should encourage scientists to continue to explore more such means to control tsetse flies and trypanosomiasis in Africa.

REFERENCES

1. BARTH, D. and SUTHERLAND, I.H.(1980). Investigations of the efficacy of ivermectin against ectoparasites in cattle. Zentralbl.Bakt.Parasit. Infect.Hyg. 1 Abt., 267, 319 n°57.
2. CENTURIER C. and BARTH, D.(1980). On the efficacy of ivermectin versus ticks (O.moubata, R.appendiculatus and A.variegatum) in cattle. Zentral-bl.Bakt.Parasit.Infekt.Hyg. 1 Abt. 267, 319, n°58.
3. BURG, R.W., MILLER, B.M., BAKER, E.E., BIRNBAUM, J., CURRIE, S.A., HARTMAN, R., KONG, X.L., MONAGHAN,R.L., OLSEN,G., PUTTER,I., TUNAC, J.B. PUTTER,I., TUNAC,J.B., WALLICK,H., STAPLEY,E.O., OIWA,R. and OMURA,S., (1979). Avermectins, a new family of potent anthelmintic agents: producing organisms and fermentation. Antimicrobic.Agents Chemother. 15, 316-367.
4. DISTELMANS,W., D'HAESELEER,F. and MORTELMANS,J. (1983). Efficacy of systemic administration of ivermectin against tse-tse flies. Ann.Soc. Belge Méd.Trop.,63, 119-125.

5. DRUMMOND,R.O., WHETSTONE,T.M. and MILLER,J.A. (1981). Control of ticks systemically with Merck MK-933, an Avermectin. Jl.Econ.Entomol. 74, 432-436.
6. EGGERTON,J.R., OSSLIND,D.A., BLAIR,L.S., EARY C.H., SUHAYDA,D.,CIFELLI, S., RICK,R.F. and CAMPBELL,W.C. (1979). Avermectins, new family of potent anthelminthic agents: efficacy of the B_1 a component. Antimicrob. Agents Chemother. 15, 372-378.
7. EGGERTON,J.R., BIRNBAUM,J., BLAIR,L.S.,CHABALA,J.C.,CONROY,J.,FISCHER, M.H., MROZIK,H., OSTLIND,D.A., WILKINS,C.A. and CAMPBELL,W.C.(1980). 22, 23-Dihydro-avermectin B_1, a new broad-spectrum antiparasitic agent. Brit.Vet.Journ. 136, 88-97.
8. JORDAN,A.M. and TREWERN,M.A.(1976).Sulphaquinoxaline in host diet as the cause of reproductive abnormalities in the tsetse fly (Glossina spp). Entom.Experim.Applic. 19, 115-129.
9. LANGLEY,P.A. and ROE,J.M.(1984). Ivermectin as a possible control agent for the tsetse fly, Glossina morsitans. Entomol.Exp.Applic. 36, 137-143.
10. MOLOO,S.K.(1985). Distribution of Glossina species in Africa. Acta Tropica, 42, 275-281.
11. MORTELMANS,J.   .Nouvelles perspectives dans le traitement antipara- sitaire. Ivermectine, produit pionier d'une nouvelle approche. In Press.
12. MURRAY,M. and BLACK,S.J. (1985). African Trypanosomiasis in cattle: working with nature's solution. Veter.Parasit. 18, 167-182.
13. NJOGU,A.R.,DOLAN,R.B., WILSON,A.J. and SAYER,P.D.(1985). Trypanotole- rance in East African Orma Boran cattle. Vet.Rec. 117 , 632-636.
14. PARENT,R. and SAMB,F.(1984). Utilisation de l'ivermectine en milieu tropical. Etude sur des jeunes bovins à l'embouche. Rev.Méd.Vét. 135, 131-134.
15. PEGRAM,R.G. and LEMCHE,J. (1985). Observations on the efficacy of iver- mectin in the control ticks in Zambia. Vet.Rec., 117, 551-554.
16. SAUNDERS,D.S. (1971). Reproductive abnormalities in the tsetse fly, Glossina morsitans orientalis Vanderplank, caused by a maternally acting toxicant in rabbit food. Bull.entom.Res., 60, 431-438.
17. TURNER,D.A. and MARASHI,M.H. (1973). A second incident of reproductive abnormalities in colonized Glossina morsitans morsitans Westw., caused by maternally acting toxicant in rabbit food. Trans.Roy.Soc.Trop.Med. Hyg. 67, 292-293.
18. VAN DEN ABBEELE,J.,GOOSSENS,M. and D'HAESELEER,F.(1986).A preliminary evaluation of the efficacy of ivermectin on the reproductive biology of Glossina palpalis palpalis (Rob.-Desv.)(Glossinidae, Diptera). Ann.Soc. Belg.Med.Trop. In press.

The ecological impact of insecticides in connection to the control of tsetse flies in Africa: A review

J.W.Everts & J.H.Koeman
Department of Toxicology, Agricultural University, Wageningen, Netherlands

Summary

In the past the chemical control of tsetse flies often gave rise to considerable and sometimes irreversible damage in the ecosystems concerned. The most dramatic side-effects were observed in situations where dieldrin and endosulfan were used at high dose rates. The methods developped over the last decade for non-residual applications of certain pyrethroids and endosulfan cause less environmental damage and in most situations recovery of affected populations of organisms can be observed within the same season.
Nevertheless, a continuing surveillance of possible environmental effects of pesticides in connection to tsetse control operations remains necessary. One may expect that the resilience of many ecosystems may decrease in future because of factors related to the steady increase of human pressure on the environment. Pesticide applications, whose side-effects appear to be of a reversible nature in relatively unspoiled habitats may cause irreversible damage in habitats, which, to some degree, have lost their recovering strength.

In the present paper an overview will be presented about environmental effects caused by pesticides used in tsetse control operations in Afrcia.
Only those insecticides will be considered, which have been applied at an operational level and whose side-effects were studied under field conditions. These are DDT, dieldrin, endosulfan, deltamethrin and permethrin. The side-effects observed will be discussed compound-wise. The type of effect and the degree of damage not only depends on the type of chemical and the dose rate, but largely also on factors like formulation, spray equipment, method of application and the skill of those who are responsible for the implementation of the control programme in one way or another. Known data on side-effects of tsetse control operations are summarised in a condensed form in table 1.

DDT

DDT is the oldest synthetic pesticide applied against tsetse flies. It has been used successfully since 1945 (e.g. Nash 1969). Because of the high succeptability of the flies to this insecticide generally low dosages (e.g. 200 g ai/ha) were applied (Matthiessen 1984). There are no records available of acute side-effects of these treatments. Higher dosages, however, appeared to cause acute death in many non-target species (Koeman et al. 1978). Although the compound is known for its property to accumulate in certain species the levels found in wild animals in connection with tsetse control operations were relatively low as compared to levels elsewhere in the world, especially in temperate regions (Koeman and Pennings 1970, Koeman et al. 1971, Koeman et al. 1978, Matthiessen 1984). This may be caused by the fact that the amounts used region-wise were relatively small as compared to amounts used in other places. It has also been suggested that especially in hot arid areas a considerable amount of DDT and

Table 1. Summary of side-effects observed after application of pesticides against tsetse flies in Africa.

G = Groundspray
H = Helicopter
FW = Fixed Wing Aircraft

R = Residual Dosage
NR = Non-residual Dosage

Type of Insecticide	Country and year	Method of Application		Side-effects observed	References
DDT	Nigeria 1968	G,	R	Acute mortality in birds, cold-blooded vertebrates, non-target insects. Residues in birds	12,14
	1976	G,	R		
	Kenya 1968	G,	R		
	Zimbabwe 1984	G,	R	Accumulation in soil and omnivorous birds	17
Dieldrin	Nigeria 1969-70	G,	R	Acute mortality in birds (especially insectivorous birds), a few mammals (bats, squirrels) cold-blooded animals	10, 12, 13
	Kenya 1968	G,	R		
	Botswana 1964	G,	R		
	Nigeria 1974-76	H,	R	Acute mortality in mammals (monkeys, squirrels), birds (especially insectivorous birds), cold-blooded animals. A wider range of birds species is affected than after residual groundspray application. Long-term effect (1 year) on abundance of insectivorous birds.	14
	Cameroons 1980	H,	R	Acute effects on non-target insects.	17
	Uganda 1961-71	G,	R	Long-term effects (1 year) on abundance of insects Residues in soil 10 years after application.	22
Endosulfan	Nigeria 1975-76	H,	R	Acute mortality in cold-blooded vertebrates (especially fish), birds (especially fly-catchers), a few mammals (bats), non-target insects. (long-term effects on insec-tivorous birds less pronounced than after similar applications of dieldrin).	1, 5, 14
	Niger 1977	H,	R		
	Burkina Faso 1978	H,	R		
	Ivory Coast 1979	H,	R	Residues in insectivorous reptiles No side-effects detected.	6, 21
	Nigeria 1976	FW,	NR	Indications for reductions in reproduction of fish	3
	Botswana 1975-77	FW,	NR		

Compound	Country	Year			Effects	Ref.
Deltamethrin	Nigeria,	1977	G,	R	Acute deleterious effects on populations of crustaceans, aquatic beetles, bugs, mayfly larvae.	21
		1978	H,	R	Acute mortality of terrestrial non-target insects,	19
	Burkina F.	1977	H,	R	especially ants. Reduction of diptera and hymenoptera.	19
	Ivory Coast	1979	H,	R	Long-term effect in one shrimp species.	7
	Burkina F.	1978	H,	R	Acute mortality in non-target insects, especially ants.	7
	Ivory Coast	1981	G,	R	Reduction of populations of muscoid flies, parasitic hymenopterans and epigeal spiders. No long-term effects.	9
	Burkina F.	1978	H,	NR	Acute mortality in crustaceans and non-target insects. No measurable reduction in populations.	5
Permethrin	Nigeria	1977	G,	R	Acute mortality in non-target insects. Reduction of mayfly larvae and dragon flies.	19
	Ivory Coast	1981	G,	R	Acute mortality in non-target insects. Reduction of populations of parasitic hymenopterans and epigeal spiders more pronounced than when deltamethrin was used.	9
	Nigeria	1978	H,	R	Acute mortality in non-target insects. Reduction of terrestrial insect populations of aquatic insect larvae and crustaceans.	19
	Burkina F.	1978	H,	NR	Acute mortality in non-target insects and crustaceans.	5

other chlorinated hydrocarbons may disappear from the ecosystems concerned through volatilisation and/or codistillation (Everaarts et al. 1971, Koeman et al. 1971). The use of DDT for tsetse control has been restricted in most countries of Africa (see also Dr. Lovemore's contribution to this meeting).

Dieldrin

In the early sixties dieldrin was introduced in tsetse control operations (e.g. Graham 1964). The most apparent side-effects of chemical tsetse control have been observed in situations where this insecticide was applied. At dosages used for residual treatments (700-900 ha) animals from a wide range of taxa have been observed to be killed (Graham 1964, Koeman et al. 1971, Koeman et al. 1978, Müller et al. 1980, Everts and Koeman 1982). A variety of species of insectivorous birds appeared to be very vulnerable (e.g. flycatchers, robinchats, and insectivorous kingfishers). Many casualties have also been recorded among fish, amphibians, reptiles and non-target arthropods. Occasionally mammals were found among the casualties including squirrels, bats, duiker antelopes and monkeys. In the Lame Burra forest reserve in Nigeria a local population of the Tantalus monkey (Cercopithecus aethiops) was virtually wiped out by a single high dose rate application of dieldrin (Koeman et al. 1978). It was found that certain species of birds were still absent or showed a very low abundance after one year. Long term ecological studies exceeding periods of two years have not been made in connection with dieldrin application for tsetse control.

Endosulfan

After its introduction in the early 70s endosulfan has been used for tsetse control in various places. The compound is less persistent than DDT and dieldrin and it is more toxic to tsetse-flies. Endosulfan is used in residual as well as in so-called non-residual applications. The former consists of a single or two application(s) at relatively high dose levels ranging from 100-1000 g ai/ha. The latter consists of repeated applications (4-5 times) at about weekly intervals at relatively low dose levels, in the order of 10-15 g ai/ha.
Mass mortality has been observed in a variety of animal species, vertebrates as well as invertebrates, after single applications at dose rates of 800 to 1000 g ai/ha., including birds (e.g. flycatchers), fruitbats, reptiles, amphibians and fish (Koeman et al. 1978, Dortland et al. 1978, Everts et al. 1978, Douthwaite et al. 1981). Cases have been reported where whole fish populations of sprayed stretches of river were virtually wiped out. Even when the forest is treated with 100 g ai/ha a considerable mortality of fish may occur (Everts et al. 1978).
There are some notable differences between dieldrin and endosulfan, especially in the time course of the mortality in non-target species. After high dose rate applications of endosulfan there was an immediate onset of deaths of birds and other vertebrates. Thereafter the numbers of casualties found on successive days gradually decreased. Almost all casualties were found within a period of five days after spraying. Mortality in insects and spiders was highest in the first two days with a marked decline afterwards. After dieldrin applications the mortality pattern showed a more prolonged effect. Vertebrate mortality continued for a period of twenty days. In insects and spiders a peak in the mortality was observed after five to nine days (Koeman et al. 1978). After non-residual (low dose) applications, apart from a slight mortality in non-target insects, no mortality has been observed in vertebrates including fish (Everts et al. 1978). However, Douthwaite et al. (1981) found some evidence for decreased reproductive success in fish (Tilapia rendalli) in areas in Botswana (Okavango delta) repeatedly treated with endosulfan at a dose rate of 9.5 g ai/ha.

Deltamethrin

Deltamethrin was the first pyrethroid used successfully against tsetseflies at an operational level (Gruvel and Taze, 1978; Spielberger et al., 1979). As a residual spray the dose levels applied amount to 12-20 g ai/ha. In repeated (5 times) non-residual applications levels as low as 0.2 g ai/ha per spray have been used. After residual applications certain crustaceans, especially shrimps appear to be highly vulnerable (Takken et al. 1978; Smies et al. 1980; Everts et al. 1983). In one case it was noticed that the severely affected large shrimp Macrobrachium raridens only showed signs of recovery two years after the application (Everts et al. 1983). The other arthropod taxa which were heavily affected recovered within the season in which the spraying took place. Although labelled as fish toxic on the basis of laboratory studies, effects on fish were not observed. Neither have acute effects been recorded in any other vertebrate species. In the terrestrial environment some species of spiders and parasitic hymenopterans virtually disappeared after ground applications at 20 g ai/ha (Everts et al., 1985). It was demonstrated, however, that of three pyrethroids tested (cypermethrin, permethrin and deltamethrin) the latter compound showed the highest selectivity towards the target-species. The biomass of phytophagous and predatory insects was only slightly reduced. Under the given circumstances deltamethrin has a half-life of 3-5 weeks (e.g. Spielberger et al., 1979). The compound is degraded relatively quickly under the influence of abiotic as well as biotic factors.

Permethrin

This is another pyrethroid which has been applied successfully in tsetse control (e.g. Spielberger et al. 1979). The side-effects observed at effective dose levels in the order of 40 g ai/ha are comparable to those caused by deltamethrin. Again no mortality was observed in fish (Everts et al. 1978; Smies et al. 1980) with one exception. At an excessively high, accidental dosage of 150 g ai/ha permethrin was found to cause fish mortality (Douben et al., 1985). In terrestrial habitats the side-effects were more pronounced than after deltamethrin applications. Of 16 major groups of arthropods 6 were significantly reduced after permethrin spraying and 4 after deltamethrin application. Species richness in an abundant parasitic hymenoptera superfamily, Proctotrupoidea, declined after permethrin but not after deltamethrin application (Everts et al. 1985).

Modification of insecticide application technology as a means to reduce the environmental impact.

The application technique by which an insecticide is applied in tsetse control may have a considerable modifying effect on the degree of impact to non-target species, both qualitatively and quantitatively. It was found for instance that aerial application in tsetse control tend to be less discriminative than ground-spray procedures. With the latter the day-time resting sites of tsetse flies can be treated selectively. By aerial applications also parts of the vegetation are contaminated which do not play a role in the life-cycle of these flies. A comparison of ground and aerial applications of dieldrin revealed some remarkable differences. Aerial application appeared to affect species which were not hit by ground spraying. Examples being a bee-eater (the Red-throated bee-eater, Melittophaga bullocki) and the Tantalus monkey. A wider range of insect species (for instance those living in the higher strata of the vegetation) may be contaminated, thus possibly affecting a wider range of predators as well (Koeman et al. 1978). The impact on the aquatic fauna may also be reduced markedly by replacing aerial application by groundspraying, supposed the latter is applied carefully enough thus causing less drift than the former method. Especially in case of residual applications one should choose for groundspray techniques when areas are concerned which are particularly vulnerable in one way or another (nature reserves, breeding and spawning sites of birds and fish respectively). There are many other aspects which are relevant, ranging from the overall spraying power of the spraying equipment to droplet size. During a spraying campaign in Niger most of the insecticide applied was not deposited in the vegetation but blown through it because the type of helicopter hired appeared too heavy for the purpose (Dortland, pers. comm.). As a consequence the fish population was wiped out, while the majority of the tsetse flies survived.

Of course the most promising modification for a reduction of damage is formed by the use of non-residual techniques as already mentioned before. In general this approach implies that an area is treated repeatedly for a total length of time which cover the full life-cycle of the tsetse fly species concerned. Repetitive treatment also implies that non-residual applicatons tend to be considerably more costly than residual techniques. However, the environmental 'costs' are much lower and this should warrant the prefered use of non-residual technology in tsetse control.

The reversibility of side-effects caused by insecticides used in tsetse control.

Non-target species, which are wiped out from a certain area, as happened for instance to certain species of birds, shrimps and fish after applications of dieldrin, deltamethrin and endosulfan respectively can only recover when a sufficient number of specimens of the same ecotype migrate into the area from other regions where stocks are still available.
Recovery may occur whithin a short period of time, when the same species is present in a nearly confluent habitat. For instance in the river in Niger when all fish were wiped out by endosulfan (Dortland et al. 1978). After 10 months sampling showed that many species had managed to repopulate the river in fair numbers. A similar observation was made in a river in Nigeria (Koeman et al. 1978). In most areas where tsetse control is practised the ecosystems are subject to regular disturbance (e.g. seasonal drought, flood) by which many populations are considerably reduced. However, they also recover normally. These ecosystems show a high resilience especially with regard to the aquatic biota. Many populations, however, depend on restricted areas (refuges) for their maintenance where minimal habitat conditions are present throughout the year (e.g. Lowe and McConnel 1975). These areas become more and more scarse and isolated. This is the result of a land-use pressure which often depasses the carrying capacity of the ecosystems. We now face a development towards an increasing pressure by chemicals while the resilience of the ecosystems is decreasing. In other words: ecological dammage caused by pesticides may become more and more irreversible.

Conclusion

At present there are no methods available for the chemical control of tsetse flies which do not have any side-effects on non-target species at all. The repeated, non-residual, low dose-rate applications of endosulfan and certain pyrethroids, however, seem to have almost negligible effects in most situations studied. These methods deserve preference over the residual techniques which generally require high dose rates.

In case of residual applications more attention should be paid to improved methodology and skill to limit drift of insecticides, for instance to adjacent aquatic habitat (viz. ground-sprays versus aerial, appropriate formulations and equipment, adequate trainings and recommitments of responsible personnel).

One should recognize the dependence of treated areas (where species may be depleted severely) on unaffected habitat elsewhere from which affected populations are restored. The acceptance of a certain degree of damage in one place should imply that efforts are made to adequately protect the threatened species in the refuges which are essential for their survival.

Literature

1. Dortland, R.J., A.C. van Elsen and J.H. Koeman (1978). Observations on side-effects of a helicopter-application of endosulfan against tsetse flies in Niger. Rep. Dept. Toxicol., Agricult. Univ. Wageningen, The Netherlands.

2. Douben, P.E.T., J.W. Everts and J.H. Koeman (1985). Side-effects of pyrethroids used to control adults of **Simulium** **damnosum** **s.l.** in a riverine forest habitat in Togo (West Africa) Rep. Dept. Toxicol., Agricult. Univ. Wageningen, The Netherlands.

3. Douthwaite, R.J., P.J. Fox, P. Matthiessen and A. Russell-Smith (1981). Environmental impact of aerosols of endosulfan, applied for tsetse fly control in the Okavango Delta, Botswana, Final Rep. Endosulfan monitoring Project, Overseas Development Administration, London, U.K..

4. Everaarts, J.M., J.H. Koeman and L. Brader (1971). Contribution a l'étude des effets sur quelques éléments de la faune sauvage des insecticides organochlorés utilisés au Tchad en culture cotonnière. Cot. Fib. Trop. **26**, 385-393.

5. Everts, J.W. G.A. Boon von Ochssee, G.A. Pak and J.H. Koeman (1978). Report on the side-effects of experimental insecticide spraying by helicopter against **Glossina** **spp.** in Upper Volta. Rep. Dept. Toxicol., Agricult. Univ. Wageningen, The Netherlands.

6. Everts, J.W., K. van Frankenhuyzen, B. Román, J. Cullen, J. Copplestone and J.H. Koeman (1983a). Observations on side-effects of endosulfan used to control tsetse in a settlement area in connection with a campaign against human sleeping sickness in Ivory Coast. Trop. Pest manage **29**, 177-182.

7. Everts, J.W., K. van Frankenhuyzen, B. Román and J.H. Koeman (1983b). Side-effects of experimental pyrethroid applications for the control of tsetse flies in a riverine forest habitat (Africa). Arch. Environ. Contam. Toxicol., 12, 91-97.

8. Everts, J.W. and J.H. Koeman (1982). Side-effects of dieldrin applications as a barrier against tsetse flies. Rep. Dept. Toxicol., Agricult. Univ. Wageningen, The Netherlands.

9. Everts, J.W., B.A. Kortenhoff, H. Hoogland, H.J. Vlug, R. Jocqué and J.H Koeman (1985). Effects on non-target terrestrial arthropods of synthetic pyrethroids used for the control of the tsetse fly (**Glossina** **spp.**) in settlement areas of the Southern Ivory Coast, Africa. Arch. Environ. Contam. Toxicol., 14, 614-650.

10. Graham, P. (1964). Destruction of birds and other wild life by dieldrex spraying against tsetse fly in Bechuanaland. Arnoldia 10, 1-4.

11. Gruvel, J. and Y. Tazé (1978). Essais d'un nouveau pyréthrinoïde: la décaméthrine (Décis: OMS 1998) contre G. tachinoides au Tchad. Rev. Elév. Méd. Vét. Pays trop. 31 193–203.

12. Koeman, J.H. and J.H. Pennings (1970). An oriental survey on the side-effects and environmental distribution of insecticides used in tsetse-control in Africa. Bull. Environ. Contam. Toxicol., 5, 164–170.

13. Koeman, J.H., H.D. Rijksen, M. Smies, B.K. Na'isa and K.J.R. Maclennan (1971). Faunal changes in a swamp habitat in Nigeria sprayed with insecticide to exterminate Glossina. Neth. J. Zool., 21, 443–463.

14. Koeman, J.H., W.M.J. den Boer, A.F. Feith, H.H. de Iongh and P.C. Spliethoff (1978). Three years' observations on side-effects of helicopter applications of insecticides used to exterminate Glossina species in Nigeria. Environ. Pollut., 15, 31–59.

15. Lowe-MacConnel, R.H. (1975). Fish Communities in Tropical Freshwaters. Longman, London, U.K..

16. Matthiessen, P. (1984). Environmental contamination with DDT in Western Zimbabwe in relation to tsetse fly control operations. DDT monitoring project, Final Report. Overseas Development Administration, London, U.K..

17. Müller, P., P. Nagel and W. Flacke (1980). Ökologischer Einfluss von Tsetsefliegenbekämpfung mit Dieldrin im Hochland von Adamaoua (Kamerun). Amazonia, 7, 31–48.

18. Nash, T.A.M. (1969). Africa's bane, the tsetse fly. Collins, London, U.K..

19. Smies, M., R.H.J. Evers, F.H.M. Peynenburg and J.H. Koeman (1980). Environmental aspects of field trials with pyrethroids to eradicate tsetse fly in Nigeria. Ecotox. Environ. Safe. 4, 114–128.

20. Spielberger, U., B.K. Na'isa, K. Koch, A. Manno, P.R. Skidmore and H.H. Couts (1979). Field trials with the synthetic pyrethroid insecticides permethrin, cypermethrin and decamethrin against Glossina in Nigeria. Bull. Entomol. Res., 69, 667–689.

21. Takken, W., F. Balk, R.C. Jansen and J.H. Koeman (1978). The experimental application of insecticides from a helicopter for the control of riverine populations of Glossina tachinoides in West Africa. VI. Observation on side-effects. Pest. Agr. News Summary 24, 455–466.

22. Sserunjoji, S.J.M. and T.C. Tjell (1971). Insecticide residues following Tsetse control in Uganda. Working paper no. 13 of a joint FAO/IAEA panel, Vienna.

Session 2
Biological, biotechnical and other control methods

Chairman: J.Mutuku Mutinga

The development of baits to survey and control tsetse flies in Zimbabwe

G.A.Vale

Tsetse and Trypanosomiasis Control Branch, Department of Veterinary Services, Zimbabwe

Summary

Field trials in Zimbabwe have shown that: 1) new traps are more effective
than ox flyrounds for surveying <u>Glossina pallidipes</u> Aust. but poor for <u>G.
morsitans morsitans</u> Westw.; 2) auto-sterilizing traps can control both species but are prohibitively complex, and 3) insecticide-coated targets can
isolate and control populations of both species but need further refinement. Present research, in cooperation with other countries, is aimed mainly at identifying more odour attractants to use with traps and targets, but
also at: 1) improving the design of traps and targets; 2) refining the insecticide formulations for targets; 3) simplifying the sterilizing devices,
and 4) elucidating optimum placement patterns for baits.

1 Introduction

There are several ways in which the further development of baits can
help to overcome the problems of surveying and controlling populations of
the savannah tsetse <u>Glossina pallidipes</u> Aust. and <u>G. morsitans morsitans</u>
Westw. Most of the present survey techniques rely on the handnet capture
of tsetse visiting mobile baits, such as an ox led by a party of men, but
such surveys are costly, requiring much labour; they catch few females and
they are poor at detecting the presence of low density populations. Traps,
provided they can be made more effective than those in common use, offer
the advantages of a low labour requirement; they work all day, every day,
they need maintenance only when in use and they catch high proportions of
females.

The common methods of tsetse control that rely on the extensive use of
insecticides require too many of the resources that are scarce in most
parts of Africa, such as ample funds, especially foreign exchange, and a
large body of trained and experienced staff to deal with the logistics,
planning and supervision of spraying operations. There is also the problem
that spraying can usually be conducted in only one season, when the weather
is suitable. This aggravates the logistical difficulties. It ensures that
planning errors cannot be corrected promptly as they emerge, and means that
tsetse are free to invade far into the sprayed territory in the months when
the spray becomes no longer effective. Moreover, there is increasing concern that the residual insecticide applied by ground based teams is ecologically damaging. Fortunately, with aerial spraying, it is possible to use
non-residual insecticides which are likely to be less damaging, but aerial
spraying cannot be performed satisfactorily in those many places where the
terrain is too broken to allow the necessarily low and straight flight at
night.

Trap methods of tsetse control are potentially useful as alternatives
to the extensive use of insecticide. The traps offer a simple technology

that can be based largely on local skills and materials; they are effective throughout the year, so avoiding the several problems of seasonal control, and they offer negligible risks of environmental damage. Such risks as may occur can be assessed by studying the few creatures other than tsetse that come to the traps - a far less daunting task than studying all of the many creatures that will contact widely broadcast sprays.

Two refinements to the trapping technique are of interest. First, the trapping operation would be about twice as efficient if a simple, cheap and safe device could be fitted to the trap to sterilize and release the flies instead of simply retaining them; the released females, being unable to re-produce, would be effectively dead and the released males would interfere with the reproduction of females that did not go through the trap. Second, if it is the intention simply to kill tsetse, it would be easier and cheap-er to replace traps by visual targets coated with insecticide. While the targets necessitate the use of insecticide, the ecological risks are very small if the insecticide is applied only to thinly spread baits that are powerful attractants for tsetse alone.

Over the last decade, several designs for targets, retaining traps and sterilizing traps have been developed in Zimbabwe. The effectiveness of these baits has been improved several-fold by the use of artificial odour attractants, justifying preliminary field trials of the baits. This paper outlines the results of the trials and indicates the further development that is now in progress and which seems necessary before the baits can be recommended confidently for widespread use.

2 Field trials

2.1 Retaining traps for surveys
Box-like traps, 90x90x90cm, made of blue and black cloth and known as F3 traps, have been deployed in the Hurungwe and Angwa/Doma districts of Zimbabwe and baited with bottles of odour attractant releasing 0.5mg/h of 1-octen-3-ol (henceforth termed octenol) and either 500mg/h of acetone or 70mg/h of butanone. Such traps have proved useful in detecting low densi-ties of G. pallidipes. For example, in one area, eigth traps operated from November 1985 to January 1986 caught a total of 27 flies of this species, as against only one caught by two ox survey teams. Since at least 20 traps can be operated for the cost of fielding one ox team, it seems that trapp-ing is much more economical and effective than the ox round for G. palli-dipes. For G. m. morsitans the traps are less effective, not because this species is unattracted to the traps but because the flies that are attrac-ted fail to enter as readily as G. pallidipes. The proportion of attracted flies that enter is about 35% for G. pallidipes but only about 15% for G. m. morsitans. Clearly, there is scope for substantial improvement in the design of traps, especially for G. m. morsitans.

2.2 Sterilizing traps
A sterilizing device has been designed which can be fitted to a trap as a substitute for the retaining cage normally used with a trap. This de-vice collects the trapped flies for 30min, allowing a reasonable number of tsetse to accumulate. It then sprays them with an aerosol of metepa, holds them in the spray for 10s and then releases them. Most of the flies go well away from the trap after release. From April to December 1981, a sterilizer was fixed to each of two to six traps, baited with 2 lit/min of carbon dio-xide and 5g/h of acetone and placed on an island of about 4.5km^2 in Lake Kariba. The island population of G. pallidipes, studied by mark and recap-ture techniques, showed a clear reduction in its birth rate in this period but no increase in its death rate. Up to 14% of the G. pallidipes females

caught by ox rounds on the island during this period showed reproductive abnormalities, such as both ovaries atrophied. Relatively few of the G. m. morsitans on the island passed through the traps, partly because this species does not enter traps readily but also because the traps were operated in the afternoon only; much of the daily activity of G. m. morsitans occurs in the morning whereas for G. pallidipes almost all of the activity is concentrated in the afternoon. Thus, the population of G. m. morsitans functioned as an internal control and, as expected, this species showed no decline in its birth rate and few reproductive abnormalities.

While the sterilizers appeared technically sound, they were extremely expensive, costing about US$1000 each. They needed much maintenance to their moving parts, required regular replacement of their batteries and had to be inspected with much care to avoid inhalation of the metepa spray. Moreover, many tsetse failed to transfer from the trap to the sterilizer, apparently because the sterilizer was too bulky, restricting the light passing through the device and so interfering with the light-seeking response necessary to take flies into the sterilizing chamber. The carbon dioxide used with the traps was beneficial in attracting tsetse from a distance and in encouraging tsetse to enter the traps, but it was far too expensive and inconvenient for routine use, requiring a pressure gauge and flow meter and a heavy gas-cylinder that needed replacement once a month.

2.3 Insecticide-coated targets

Many of the tsetse that visit targets fail to alight on them and so do not contact the insecticide deposited on the target. This problem can be overcome by placing a sheet of fine black netting beside the target so that tsetse collide with the net while flying round the target, picking up insecticide deposited on the net. The targets used for the field trials incorporated this refinement. Each consisted of a sheet of black cotton cloth, about $1m^2$, as the visual target, fixed to a sheet of terylene netting of about $0.5m^2$. A white roof of plastic-coated taffeta, about $0.5m^2$, covered the netting to protect the insecticide deposit from vertically-falling rain and sunlight. The whole device was hinged on a steel post to swing like a wind vane, keeping the netting downwind behind the cloth which was mounted cross-wind to shelter the netting from driving rain.

Twenty of these targets, coated with deltamethrin suspension concentrate and baited with 100mg/h of acetone and 0.5mg/h of octenol, were placed evenly over the Kariba island in May 1983 when mark/recapture studies indicated that about 2000 G. m. morsitans and a few hundred G. pallidipes were present. The sterilizing traps were absent at this time. After placing the targets, the numbers of both tsetse populations declined rapidly. No G. pallidipes could be detected after 11 weeks and no G. m. morsitans after 9 months, despite surveys that were about 100 times as intense as those normally performed to detect tsetse elsewhere in Zimbabwe. It was known from earlier work that the island provided an excellent habitat for the flies, allowing their populations to increase 8-fold per year in the absence of controlling baits. Thus, it appears that the targets can quickly eliminate tsetse populations, even those that have a high potential for increase.

While the island work showed that the targets could deal adequately with an isolated population, it was necessary to determine how they would perform in the more usual situation in which the treated area would be subject to invasion from untreated areas nearby. An area of $600km^2$, known as the Rifa and situated in the Zambezi Valley near Chirundu, was chosen for this study. The area formed an equilateral triangle, with one side represented by the Zambezi River, another by the steep rise of the Zambezi Escarpment and the other being the Harare/Chirundu road. Dense populations

of G. pallidipes and G. m. morsitans occurred in the Rifa in association
with an abundance of wild hosts. The area was particularly suited to studies of invasion because the expected pressure of invasion was focused; few
flies occurred across the Zambezi in Zambia and so little invasion was expected from this direction; the flies in the escarpment were to be controlled by ground spraying, so eliminating that area as a source of heavy invasion. Thus, the main invasion pressure could be expected from the dense
populations across the road.

From March to May 1984, 1200 targets were placed at 100m intervals
around the border of the Rifa, and in June and July another 1200 were put
at 150-300m intervals along tracks inside the Rifa, the tracks forming a
network with a mesh of about 3x6km. In September and October another 500
targets were placed in groups of about five in areas within the network,
the targets within groups being 150-250m apart. The catches from a variety
of survey techniques in the Rifa, expressed as a proportion of control catches 25km away across the road, began to decline rapidly for both species
from April onwards. The mean age of tsetse caught in the Rifa declined,
consistent with the presumed increase in death rates. In February to June
1985, 11-15 months after the first targets had been positioned, no tsetse
of either species was caught in the middle of the Rifa, despite intensive
surveys and despite the continued abundance of tsetse in the control area.
A few tsetse were caught towards the edge of the Rifa, most being near the
road, indicating that the road was indeed the main invasion front.

In February to June 1985, surveys from 15km inside the Rifa to 15km
outside, spanning the road, indicated that the abundance of tsetse was reduced for about 7km outside of the target-treated area. Alongside the road
the population density was down by about 90%, and at 3-5km inside the Rifa
it was down by about 99.99%, tsetse being undetectable further inside. This
indicates that the targets were able to stem invasion, partly by killing
tsetse before they moved far into the Rifa, and partly by reducing the invasion pressure on the border.

Tabanidae and Muscoids were the only insects other than tsetse that
appeared consistently at the targets. The population densities of these
creatures showed no changes due to the targets, presumably because their
attraction to the targets was relatively weak or because their lost numbers
were readily replaced by a reproductive capacity greater than that for tsetse.

The practical experience of erecting and maintaining targets over a
large area was instructive. The targets along the tracks were easy to erect
and maintain, but those between the tracks were more troublesome because
much walking was necessary to deal with each small group of them. Since the
tsetse populations were declining sharply before the placement of the targets between the tracks, it might have been better to have avoided placing
these targets and to be willing to wait a little longer for control to be
achieved. During the wet season it was necessary to clear the grass around
the targets, to preserve their visibility, to allow them to turn properly,
and to form a fire break to prevent their being burnt in the dry season.
This clearing was performed at two to three month intervals, when the targets were visited for repair, respraying and replenishment of attractants,
but the clearing took twice as much effort as the all of the other operations together. It might have been advantageous to have sprinkled weed-
killing pellets when the targets were erected. Surprisingly few of the targets were damaged by wild animals; an average of less than 1% per month
needed major repair or replacement, mostly due to rhino attack.

From July 1985 onwards the targets started to become noticeably weathered, their roofs decaying, their nets becoming torn and the black dye

on the cloth and net fading so that the targets became light grey. In October 1985, the catch from electrified versions of weathered targets that had been exposed in the Rifa since March 1984 indicated that the targets were only about a third as effective as when they were new. This was largely because of their fading since the effectiveness of the targets doubled when they were redyed. Seemingly coupled with the reduced efficacy of the Rifa targets, the abundance of tsetse in the area began to increase slightly from about September 1985. However, even in February 1986, tsetse were exceedingly scarce or absent over much of the Rifa and there had been no great invasion across the road. The tsetse situation in the Rifa now, after two years of target operations, is no different from that which could reasonably be expected if the area had been sprayed with insecticide in 1984 and 1985, especially in view of the heavy pressure of invasion on the one side. The costs of the target operation were about equal to the expected costs of spraying with residual insecticide, which is the cheapest spraying technique available.

3 Further development

3.1 Odour attractants

The field trials showed that each of the various baits has its own particular need of further development. However, all baits share the need to be more attractive, especially since increased attractiveness will minimize their other problems. Thus, if the attractiveness of baits is increased, fewer will be required and the unit cost of the baits and the difficulties of deploying and servicing each one become less important. It is encouraging, therefore, that host odours are known to contain attractants which, although not fully identified, are able to increase catches by at least ten times when added to the artificial attractants already in use. Attempts to identify new attractants are being made by cooperative research involving laboratory studies at the Tropical Development and Research Institute, London, and the Tsetse Research Laboratory, Bristol, and field studies by the Department of Veterinary Services, Zimbabwe, and French and German institutes in the Ivory Coast, Burkina Faso and the Congo.

A particularly promising chemical, termed Omega and known to pass through filters of charcoal and soda lime, is present in host breath. This very volatile substance seems unattractive itself but it does improve the effectiveness of the other known attractants, increasing the numbers of tsetse attracted from a distance. Omega appears important in the production of the linear dose/response relationship that is evident with natural host odour, as against rapidly-levelling dose/response curve that occurs with carbon dioxide, acetone and octenol. Other, much less volatile attractants are known to occur on the sacks on which bushpigs have slept and in the urine of cattle. The attractants in these materials may lure tsetse from a distance, but one of their clearest effects is to encourage tsetse to enter a trap after arriving near it. This suggests that the attractants could be economical substitutes for carbon dioxide in improving trap performance.

So far, most attention has been directed at identifying the attractants on the sacks and in the urine. At least four attractants seem present in these materials, at least one being polar and at least one being non-polar, with none of the main attractants being destroyed or strongly bound by acid or base. Much of the activity is in the phenolic fraction; cresols have proved attractive against G. m. submorsitans in Burkina Faso but ineffective so far against G. pallidipes and G. m. morsitans in Zimbabwe. A promising substance, of molecular weight about 208 and consisting of only carbon, hydrogen and oxygen, occurs in the non-acidic, non-aromatic group.

Field studies are being made of the responses induced by known attractants at various distances from the source of attractant and in the presence and absence of various visual stimuli. This work should assist the refinement of laboratory techniques for screening candidate attractants. It may also suggest ways of using the known attractants more effectively.

3.2 Trap design

Several new traps have been designed which are about twice as effective as the F3 for G. m. morsitans and G. pallidipes but the new traps are about twice as large as the F3 and so are not suitable for routine use. A reduction in the size of the big traps developed so far has reduced their effectiveness. The F3 is being refined by making its frame light and collapsible and by providing the trap with a cheap retaining cage which, unlike the netting chamber now in use, cannot be eaten by grasshoppers and which delivers trapped insects to a special storage container protected from ants and rain. This should allow the traps to be visited at less frequent intervals, with less fear of the cages becoming clogged or damaged and the catch becoming rotten or destroyed.

3.3 Target design

Following the demonstration that deposits of deltamethrin suspension concentrate are resistant to being washed away by rain and are not destroyed rapidly by sunlight, a roof as on the Rifa targets was considered unnecessary. A number of new, roofless targets have been designed and one of them, termed the S type, is about twice as effective as the Rifa target and about half the cost. It consists of a 1m square screen of black cloth with a 1x0.5m sheet of black net extending out from each vertical edge of the screen. Current experiments suggest that the alighting responses of tsetse can be increased considerably by changing the shape and size of the screen so that the netting may be unnecessary. The avoidance of nets would make the targets cheaper and more durable. Further refinements to the design of traps and targets are expected to follow from the present study of video recordings of tsetse near such objects.

3.4 Insecticide formulation and dyes

A water soluble chemical that absorbs ultra violet light (2-hydroxy, 4-methoxy benzophenone 5-sulphonic acid), used as a 0.1% additive to a 0.1% suspension of deltamethrin, has been found to protect the insecticide from photo degradation. It prolongs the effective life of the insecticide deposits from two to three months to four to five months in dry weather. Studies with stickers to improve resistance to rain washing are in progress. Black dyes (Colanyl and Duasyn) at up to 2% in the insecticide spray do not affect the efficacy of the insecticide. This suggests the possibility of rejuvenating the blackness of the targets while respraying with deltamethrin. The dyes are needed at only 1% for treating a completely faded target and at only about 0.1% as a repeated treatment to prevent fading. The ultra violet absorber is also likely to retard fading.

3.5 Sterilizer design

Attempts are being made to develop a sterilizer that involves the exposure of tsetse to an exceedingly dilute vapour of bisazir as they find their way slowly through an open plastic bottle fixed to a trap. Such a device would be much cheaper, simpler, less bulky and less hazardous than the sterilizer used on the island. Even if a satisfactory sterilizer can be produced, sterilizing traps are likely to be more costly and more complex than targets. However, such traps could be particularly advantageous in areas where few tracks exist. Sterilized males released from traps at a few accessible places would enter the trackless regions to deal with the females there, avoiding the need for staff to enter. Also, it is worth considering the possibility of using the traps to catch male tsetse in areas of high fly density, followed by sterilizing the flies and transporting them for release in other areas where tsetse are to be controlled. This technique may well have many problems, such as the fact that the released flies could introduce new stains of trypanosomes to the treated area, but the "farming" of sterile males could be cheaper and simpler than the rearing operations now being conducted in several parts of Africa.

3.6 Population studies

Studies are being made of the population dynamics of tsetse, the distribution of the flies, their rates of dispersal and their availability to baits at various age and at different stages of the hunger cycle. This will allow a fuller understanding of those placement patterns and densities of baits that will be the most economical and effective for eliminating tsetse and providing invasion barriers.

4 Discussion

The field trials of sterilizing traps indicate that these machines cannot be used routinely in their present form. However, the present froms of survey traps and targets do seem suitable for use now, provided their limitations are appreciated and care is taken to explore their applicability before deploying them on a large scale in new types of situation. The trials of the targets are especially exciting, demonstrating the possibility of using a cheap, simple and ecologically safe method to clear an infested area and, equally important, to combat invasion.

There is ample scope for improving all of the sorts of baits described here, to produce devices that are much more economical and effective and which can be recommended confidently for use in a wider range of situations against a greater number of tsetse species. In the latter regard, it is encouraging that traps and insecticide-coated screens have been developed for riverine species in West and Central Africa, but disappointing that effective odour attractants other than carbon dioxide have not been discovered for these flies. However, it was shown long ago by C.J. Persoons that washings of pig skin were attractive to riverine tsetse near Lake Victoria, so that the search for attractants for riverine flies should not be abandoned.

The potential importance of improved baits in all of the tsetse belts justifies continued support for research into bait improvement, and demands that the outcome of this research be circulated rapidly and freely in two-way cooperation that avoids exclusive reliance on the tardy publication of results and ideas in scientific journals. There is a large and growing group of scientists interested in such cooperation, especially in the field of odour attractants. However, establishing the machinery for widespread cooperation is difficult, particularly in view of the universal wish to side-step regimented obligations, lengthy correspondence and excessive travel. Fortunately, there are usually one or two general conferences each year to which most tsetse research units are able to send representatives; it would be useful to ensure that persons interested in baits had a side meeting at such conferences, where the research could be discussed and integrated with a minimum of expense and bureaucracy. Dr. A.M. Jordan (Tsetse Research Laboratory), an attender at most tsetse conferences, has volunteered to arrange such side meetings whenever appropriate. One meeting has already been held alongside the Trypanosomiasis Seminar in Salford in 1985. At this meeting Dr. D.R. Hall (Tropical Development and Research Institute) undertook to provide, on request, news and samples of candidate attractants for field testing.

5 Acknowledgements

The author has relied heavily on personal communications of the research of Prof. E. Bursell, Messrs D. Lovemore, S. Flint, S. Torr and J. Lancien, Drs. D. Hall, M. Hall, H. Politzar, P. Merot, W. Kupper, J. Hargrove, G. Cockbill, A. Challier, C. Laveissiere, P. Langley, G. Gibson, J. Brady and M. Warnes. The research in Zimbabwe has benefited from funds or equipment donated by the European Economic Community, the Overseas Development Administration of the UK, the International Atomic Energy Agency, the Food and Agriculture Organisation, the World Health Organisation, the Wellcome Foundation Ltd and Cooper Zimbabwe Ltd. The impetus to investigate urine was provided by the publications of Dr. M. Owaga.

Trapping committed to rural communities for the control of sleeping sickness

C.Laveissière
Institut Pierre Richet/OCCGE, Bouaké, Ivory Coast

SUMMARY

Traditional techniques for the control of human-trypanosomiasis vectors are not suitable for use in the forested regions of West Africa because environmental conditions make them too difficult to use and/or too expensive. A pilot project has, however, shown that traps can be used to reduce G. palpalis populations, provided that the traps are deployed by those rural communities exposed to the disease within an endemic focus. Using this system, a large area can be covered in a very short time and with spectacular results that are immediately evident to the population. Moreover, it ensures proper care for the control equipment, which becomes the property of individual local farmers and is therefore maintained in such a way as to ensure its effectiveness. There is, however, another invaluable benefit of mobilizing rural communities: it encourages an exceptional level of attendance during visits by medical personnel and hence the rapid sterilization of the disease's human reservoir.

1. INTRODUCTION

African human trypanosomiasis attracted renewed attention in West Africa a few years ago, not so much because of the damage it causes - although this is significant - as because of the difficulties encountered in controlling the disease, particularly in forested areas. Indeed, everything favours the continued existence of this fatal endemic disease: the ubiquitous tsetse flies which colonise all facies of the forest, the inaccessibility of the population (of which three quarters may be living in compounds near their crops), the mobility of labourers and migrant planters, the difficulty of indentifying the trypanosome in suspected cases and, finally, the shortage of funds.

2. THE CHOICE OF METHOD

The control of sleeping sickness in forested areas demands both an attack on the vectors and exhaustive screening of those suffering from the disease, in order to sterilize the human reservoir. Steps must be taken as quickly as possible to eliminate all insect carriers of trypanosomes, to prevent teneral adult flies becoming infected by taking their first bites from a sick person, to maintain numbers at a reduced level for as long as possible, given the presence of possible animal reservoirs (wild or domestic), and finally, to carry out strict medical surveillance.

Trials, carried out by the Organization for Coordination in the Control of Major Endemic Diseases in West Africa and by WHO, in the foci of Vavoua and Bouaflé, have shown that residual sprays, whether administrated at ground level or from the air, are either too difficult to apply or ineffective. By contrast, trapping using deltamethrin-impregnated screens has proved to be a rapid and easy technique, but of only average effectiveness where a single application was made in only a small area. This handicap could, therefore, be removed by installing screens or traps throughout a focus and, if possible, expanding their use to the endemic zone. In its turn, this attractive solution gives rise to a major problem: how can thousand of hectares be treated by a small team and yet in accordance with the deadlines for cooperation with medical teams. On the basis that only a planter knows his own plantation, it was absolutely logical to consider entrusting him with the control equipment so that he could himself treat his own property.

The major part of the control programme concerned the plantations, which are the highest risk zones, in the hope that tsetse habitats of lesser importance - such as clumps of forest, fallow ground and fields - would be flushed out indirectly. This left possible tsetse reservoirs (the outskirts of villages and gallery forest) and facies along which reinfestation could occur (roads). Since these sites are on public land, and therefore not suitable for treatment by rural communities, the OCCMED team carried out this work using biconical traps in gallery forest and residual deltamethrin sprays 12 g/km on the edges of roads and villages.

3. MOBILIZATION OF RURAL COMMUNITIES

Before the planters of the Vavoua focus could be mobilized, it was necessary to make them aware of the dangers of the disease, to make them understand the advantages of combatting it and to teach them to handle the screens . Meetings at village level showed that most people knew of the disease because a relation or friend had been infected by, or even died of it; the tsetse fly itself is also known, not as a vector but as a pest in the plantations. The demonstration of its role in transmetting the disease comes as news to the local farmers, who rapidly realize the risks they run. At that point, their cooperation has already been gained and the equipment can be handed out, once the necessary explanations about installation and maintenance have been given. In November 1983, 8 592 hectares were protected by 15 592 blue screens (impregnated with 150 mg of deltamethrin), which had been distributed to 363 planters (representing a population of approximately 8 000 people) for the treatment of 451 coffee and cocoa plantations. At the same time, 108 km of land along roadways and around villages were sprayed and 95 biconical traps placed in 10 km of gallery forest. The operation lasted less than six days. Further meetings were held with the planters, three and six months after the beginning of the campaign, to provide them with the necessary insecticide to reimpregnate their screens: at least 90 % attended these meetings, the rest having returned to their home villages.

A gross decrease of 91 % in G. palpalis populations was achieved within a week, this figure stabilizing at 98 % for 5 months and then at 85 % for succeeding months.

These very satisfactory entomological results were associated with an exceptional level of attendance at medical examinations.

4. ACCEPTANCE RATHER THAN TOLERANCE OF MEDICAL EXAMINATIONS

Having become aware of the risks involved, the role they could play and of how their own interest were at stake, the local people attended medical inspections in great numbers. Whereas even under the most favourable conditions, routine surveys reached no more than 40 % of the population, the increased awareness on the part of local farmers raised average attendances to 84 %. In Koetenga, the village most severly affected by the endemic disease since 1975 and the one most extensively involved in control measures, 98.2 % of villagers attended inspections!

Under this circumstances, it is reasonable to assume that the screening for cases of the disease was exhaustive.

5. COST OF THE OPERATION

Because of insufficient funds for its continuation, it was unfortunately necessary to terminate the pilot project in the Vavoua focus after just one year. Nevertheless, it is possible to draw up a balance sheet which could serve to determine the budget for a similar campaign.

The cost per hectare of protection is CFA 1 938 during the first year, of which 82 % comprised equipment. Insecticides (four reimpregnations) account for less than 10 %. In subsequent years, the cost falls to CFA 300, assuming a 10 % write-off of equipment.

The expenditure per inhabitant is CFA 2 000 for the first year and CFA 250 for subsequent years.

Quite satisfactory results were obtained at a relatively modest cost, but even better results will soon be possible.

6. FUTURE PROSPECTS

A number of major points emerged from the Vavoua pilot project:

- current screen design is still too inefficient;
- this is primairly because the currently-used insecticide does not remain long enough on the screen;
- ineffective equipment means that a greater number of decoys must be used for any given area. This not only raises costs appreciably, but also hinders the planter's work.

There are grounds to believe that work being conducted at the Institut Pierre Richet/OCCGE will soon eliminate these handicaps.

The Commission of the European Communities is financing a research programme aimed at improving trapping/control and trapping/sampling methods. A screen that attracts at least 40 times the number of tsetse flies has already been developed.

For three years now, the Special UNDP/World Bank/WHO (WHO/TDR) programme has been funding research into the optimal combination of the insecticide, its formula, its dosage and the kind of textile substrate used: some combinations retain their effectiveness for more than four months. Currently-used screens cannot be expected to do so for more than two months.

Finally, WHO/TDR has allocated us the necessary funds for research into an olfactory attractant capable of attracting glossina to the poisoned decoy (a technique already used in Zimbabwe).

As traps are made more efficient/attractive and impregnation can be carried out less frequently, it should prove possible to:
1) reduce the number of screens per hectare,
2) reduce the cost of a campaign,
3) further improve participation by planters.

7. CONCLUSION

Control measures against trypanosomiasis, wether human or animal, have made considerable progress in recent years, thanks to the use of traps. The virtues of these are no longer in doubt: they are cheap, immediately effective, quick to deplay and of proven harmlessness to the environment. As far as human health is concerned, their use by rural communities exposed to the disease is a further step forward since not only is it possible to achieve quick coverage of huge areas but also, by motivating population, one can achieve better attendance at medical examinations. Nevertheless, much remains to be done: before entomologists can further perfect their techniques, there will have to be even more effective participation by parasitologists in the struggle against sleeping sickness. It is difficult to see how tsetse flies can be eradicated completely, particularly in forested areas. Moreover, whatever technique is used, the flies will sooner or later reinvade their habitats. Can one therefore hope for the disappearance of sleeping sickness if it remains impossible to screen all cases and if there is no rapid solution to the problem of the naimal reservoir?

Integrated control against the Glossines in Burkina Faso

I.Kabore & B.Bauer
Centre de Recherches sur les Trypanosomoses Animales, Bobo-Diolasso, Burkina Faso

SUMMARY

The I.E.M.V.T./G.T.Z. Centre for Research on Animal Trypanosomiasis at Bobo-Dioulasso has carried out a control programme for three tsetse fly species in a 3500 km^2 area of seasonal grazing th the south of the town.

This campaign, which involved the integrated use of sterile males and insecticide-impregnated screens, was carried out in two phases between 1981 and 1984:
- during the first (preparatory) phase, a network of tracks was created in the area, entomological studies of the area were carried out and mass-rearing was developed in the laboratory for the following three tsetse fly species: G.p. gambiensis, G. tachinoides and G.m. submorsitans.
- during the second phase, that of actual control measures, the following two methods were alternated according to the season: insecticide-impregnated screens were set out in the dry season, while major releases of sterile males were carried out during the rainy season.

Enormous progress has been made in tsetse-rearing techniques, thus permitting large-scale use of the sterile male technique and at a cost competitive with that of other methods.

1. INTRODUCTION

Research on the biological control of tsetse flies has intensified in Africa in recent years. Major production units, where the flies are fed either on host animals or partly or totally on artificial membranes, have been established and have thus allowed field use of the sterile male technique. Among campaigns of this nature, one can mention those carried out in Tanzania (Tanga) and Nigeria (the BICOT project at Vom).

In Burkina Faso, this advanced technique had already been succesfully applied against Glossina palpalis gambiensis over an area of 100 km^2 before being applied on a large scale, in a pastoral zone measuring 3500 km^2, and integrated with the use of insecticide-impregnated screens. Almost 1 000 000 sterile males could be released, thanks to the development of mass-rearing of tsetse flies in the Bobo-Dioulasso laboratories.

2. LABORATORY WORK : REARING OF GLOSSINA

Laboratory research, which began in 1979, became more intensive in 1982 with the construction of two new insectaria and improvements to the former insectarium III. This research has led to the rationalization of rearing techniques for the three tsetse fly species present in the project zone – Glossina palpalis gambiensis, Glossina Tachinoides (riverine tsetse fly) and Glossina morsitans submorsitans (savannah tsetse fly).

With all of these species, host animals could gradually be replaced by artificial feeding methods using cow or pig blood from the local municipal abattoir.

The blood is defibrinated or heparined (4 I.U./ml) and irradiated (50–55 krad) before being fed to the flies. The blood is irradiated to prevent any bacterial infection and any development of trypanosomiasis in the tsetse flies. Statistics indicate that 15 to 20 % of animals sent to the abattoir carry trypanosomiasis. Freeze-dried cow and pig blood was tested but then abandoned because of its poor performance with respect to fresh blood and because of the very high cost of blood treated in this way. At the same time as the development of artificial feeding, CRTA research made it possible to simplify the storage and handling of tsetse flies. In particular, this involved the use of large rearing cages, large heating plates made of local materials and better-designed storage trolleys.

Together, these technical improvements enabled a much more rapid growth of the various colonies, which achieved a record total in 1984 of 330 000 reproductive females.

2.1. G.p. gambiensis

The colonies of this species were deliberately limited to approximately 150 000 females of which 100 000 were kept in large cages (39 x 9 x 6 cm) in groups of between 100 and 300 per cage. After 4 days feeding on rabbits, the flies were fed on heparined cow's blood for a further 6 to 7 days. The production unit could be run by only 4 people, one person for 20 – 25 000 flies.

2.2. G. tachinoides

Up to 100 000 females of this species were kept in "camembert" cages, 20 – 100 per cage, and fed on membranes using defibrinated cow's blood. Another colony of 14 000 flies was fed partly on a membrane using defibrinated cow's blood and partly on rabbits.

2.3. G.m. submorsitans

The total size of the morsitans colonies did not exceed 40 000 females. Kept 15 to a cage, these were fed exclusively on membranes. During the first 4 to 7 days, defibrinated cow's blood was used and defibrinated pig's blood during the subsequent 2 to 7 days.

At the end of the tsetse-control programme at Sidéradougou, stocks of the various species were significantly scaled down in December 1984.

3. PREPARATORY WORK FOR THE GLOSSINA-CONTROL PROGRAMME

The works required in the area before the actual control programme could begin included in particular:
- the creation and maintenance of a network of tracks (totalling 600 km) to provide access to the entire drainage system of the zone.

- an entomological survey carried out over the entire drainage system, where by a biconical Challier Laveissière trap was deployed every 100 m for one full day. In the Savannah area, parallel transects were carried out every 4 to 5 km. On the basis of this survey, a 1:50 000 scale mape could be drawn up of the distribution of tsetse flies in the Sideradougou area.
- the establishment and maintenance of a system to isolate the zone. The entire drainage system was shielded by three permanent barriers comprising traps and screens – a particularly strong one (1 700 traps and screens) against G.m. submorsitans being deployed in the south-east sector.

4. THE TSETSE-CONTROL PROGRAMME AT SIDÉRADOUGOU

This programme comprised the alternate use of insecticide-impregnated screens during the dry season and the release of irradiated males during the rainy season.

4.1. The deployment of insecticide-impregnated screens

As soon as the river survey had been completed, and before the sterile males were released, a start was made on deploying the insecticide-impregnated screens. These screens are 1 m^2 pieces of blue cloth soaked in deltamethrin (200 mg/screen). They were sited every 100 m, hung from tree branches along river courses or placed on metal frames in hollows and in savannah areas. The spacing between screens was reduced to 20 to 30 m in dense forest. Alternatively, they were sometimes interspersed with insecticide traps. In 1983, two months after the siting of 7 204 screens, the tsetse flies density had fallen by an average of 92 % for G. tachinoides and 88 % for G.p. gambiensis.

4.2. Release of sterile males

Following removal of the screens, the first releases were carried out on 19 May 1983. Releases continued without interruption until 31 December 1984, except for some areas where the screens were replaced for a two-month period.

The flies were marked, fed and then irradiated (10 000 rads for G. tachinoides, 11 000 rads for G.p. gambiensis and 12 000 for G.m. submorsitans), before being transported in containers lined with wet jute fabric designed to provide the temperature and humidity conditions required by the flies.

Releases were carried out every two weeks at each release point, these being 2 km apart and identified with coloured plastic markers.

In total, more than 900 000 irradiated males (713 382 G.p. gambiensis and 225 106 G. tachinoides) were reared at Bobo and released in the field. Over the whole area, quite acceptable ratios of sterile males to indigenous males were achieved: 13.5:1 on average for G.p. gambiensis and 20:1 for G. tachinoides. The integrated programme, involving the release of sterile males and the use of insecticide-impregnated screens, completely eradicated tsetse flies from the Sidéradougou area at a cost competitive with other methods (see F.E. Brandl's report).

5. CONCLUSIONS

The tsetse control programme in the Sideradougou pastoral zone was thus based on alternate use of two new non-polluting techniques.

The creation in the field of a network of tracks make it possible to carry out an entomological survey of the zone and the 1:50 000 scale mapping of the respective densities of the three tsetse fly species present.

Moreover, laboratory-based technical advances enabled the establishment of a mass-rearing unit (330 000 reproductive females), the first of its kind in Africa and totally independent, using techniques adapted to African conditions. The sterile male technique, associated with the deployment of insecticide-impregnated screens, thus led to the elimination of tsetse flies in a 3 500 m^2 km pastoral zone undergoing thorough development.

Considerations about the control of tsetse fly in Mozambique

J.A.Travassos Santos Dias
Instituto de Investigação Científica Tropical, Lisboa, Portugal

Summary

The Author analyses the situation of trypanosomiasis in Mozambique and refers its importance to the livestock production due
to the presence of tsetse-fly in 2/3 of that country. After re-
vising the extention, continuity and weak vulnerability of the
more important fly-belts common both to Mozambique and several
neighbouring territories, the Author considers some relevant
and urgent questions to determine where, under what circunstan-
ces, how and when actions must be taken aiming the land recuperation and minimizing the threat represented by the tsetse species, mainly G. morsitans and G. pallidipes.

Introduction

Among the territories of Eastern and Southern Africa, Mozambique constitutes undoubtedly the one where tsetse fly shows the
bigest threat, preventing the cattle-breeding in about 2/3   of
respective area.
    As a matter of fact, if one will study a map of that country
where the tsetse fly distribution is shown, easily can see that,
excepting some areas of relative reduced extention (revealed by
certain plateau zones), the glossinic communities – with G. morsitans and G. pallidipes as predominant species – spread continuously from Rovuma river in the North, down to the 22º parallel
slightly to the south of Save river.
    If, as a consequence from the occurrence of both species,the
maintenance of endemic foci of sleeping sickness had been confir
med in the provinces of Cabo. Delgado, Niassa and Tete, such cir
cunstances had not been, however, impeditive to the stablishe –
ment of human communities, though some signs of their diminishing density can be seen in one or another zone. Nevertheless,
rhe same does not apply to the cattle-breeding, in spite of many
areas ravaged by the tsettes flies, who can offfer excellent a-
grological conditions for pasture.
    As this situation goes back a very long time, it can easily
explain why in the majority of the population groups in   that

country (exceptiing those situated in the provinces of Tete, I
nhambane, Gaza and Maputo) lacks a traditional pasture activi
ty.

Whereas, the study of the oresent behaviour of tsetse flies
communities with great epidemiological importance allow us to
conclude, on their apparent stability in the majority of respec
tive range areas, the same can not be said from the southern
borders,where range expansions and contractions are too much e
vident.

In reference, it is important to recall here an historical
survey carried out by us in 1962, where we concluded that, in
the end of kast century both G. morsitans and G. pallidipes had
a more southern range, spreding in areas till the south of Lim
popo river, already belonging to the hydrographical basin of
the Incomati river.

As it is known, that situation was completely changed soon
afterwards the occurrence of the rinderpest, which by the prac
tical elimination of normal sources of nourishment of the flies
as well by their infection with the virus of that disease, com
pelled the glossinic communities restricted to savana biotopes
and afected by the pet morbus, to suffer a deep diminishing den
sity, in such strong way that in practice the whole area south
of the Save river became almost free of the tsetse flies.

However, it became quite clear that soon after the suffered
mischance, the tsetse flies (specially G. morsitans) once restored their traditional food sources, through the restablish-
ment of herds of ungulates, immediately began to spread over
the anterior loosed areas, in such way that few decades later
it had again crossed the Save river on its southwards advance.

It is commonly known that when HORNBY, in 1946, carried out
a survey of the western region of Mozambique southwards of the
Save river,.although recording a small spot of G. morsitans
near Massangena, clearly stated on the basis of his wide knowledge of the tsetse flies ecology and ethology, that "Fly popu
lation in this southward bulge of the great morsitans belt which
finds its full habitat north of the river Save is not great e-
nough to permit a very wide spread of the flies during one rai
ny season. Most dry seasons are sufficiently severe to cause
these pioneers to perish or fall back again to the Save."

Unfortunately, the reliable knowledge we have that G. morsitans and G. pallidipes had anteriorly spread all over the lar
ge area between Save and Incomati rivers, whose ecologic condi
tions should be still unchanged, thus providing adequate ambi-
ent for the survival of those flies, make it impossible for us
to share the optimism shown by HORNBY.

This opinion is also shared by a considerable number of portuguese, rhodesian and south african entomologists who dedica--
ted themselves to the study of that problem, whose studies re-

sulted in the establishment of plans and field actions to res
train the advance of the flies, specially G. morsitans, with
particular incidence on the so called "bulge of Massangena".

It is important to add, for settling definitively the situa
tion of the tsetse fly on the south of Save river, that as a
result os several surveys carried out on the area for a considerable number of years, it was possible to grant the advance
of G. morsitans along the right bank of that river, as well to
record the fact that G. pallidipes has also practically settled
on the biotopes previously occupied by the other species.

These are, therefore, the major problem that Mozambique has
to face due to the presence of the tsetse fly:

a) The existence of large areas of grazing land practically
closed to cattle breeding, which activity could be the best
way to promote the development of those vast regions.

b) A permanent and ever increasing threat of invasion of the
southern areas of Mozambique, south of the Limpopo river, where the main cattle ranches in the country are to be found. As
HORNBY said in 1946, if that were to happen, it "would be a di
saster of the first magnitude to Mozambique."

Description

Having thus stated the problem, some considerations are in
our opinion deeply important for it resolution, particularly re
garding the control of the tsetse fly.

Due to extent and continuity of the main tsetse belts into
the neighbouring territories it is important that any action
intended to promote the control of the referred species requires a combined effort by the countries affected by the problem.
The struggle to be undertaken must take in consideration not
only the cattle herds presently in danger, but also the potential capacity of the tsetse to occupy zones which, so far,have
been considered free of the fly but which have ecological conditions favourable to its existence, allowing it to come dange
rously close to the areas of animal breeding.

Based on these assumptions we are of the opinion that prio--
rity should be given to the following actions for the control
of the tsetse flies in Mozambique:

a) Detection of the progress of the tsetse flies in the
southern bulges, below the Save river, which should take the
form of a combined action of the three countries threatned by
the advance of the fly, namely Mozambique, South Africa and
Zimbabwe.

So far we have only discussed the reason for the need of
such and action on Mozambique, but it is also important to remember that in Zimbabwe there is also an important G. morsitans

belt, south of the Save river (called "Southeast belt"), which
occupies practically the whole of the "Gonarhezou Reserve".The
refore, to the "bulge of Massangena" shown in the enclosed map
- partially obtained from a report by LOVEMORE - as well as to
the "bulge of Muabsa" (although situated far from the Limpopo
river), on should add the "Gonarhezou belt", which has already
established contact with the "Massangena belt". With regards to
South Africa, the danger seems even greater than in Mozambique,
due to the fact that the Limpopo river is the northern limit
of the well knon "Krueger National Park", which, due to type of
vegetation and the number and diversity of its   animal life,
could act as match in the dissemination of the fly, placing in
danger the cattle ranches both to the West and to the East (the
se obviously in Mozambique).

The cooperation which we deem necessary must take into con
sideration to avoid the tragic economical consequences which
could result if the threat of the tsetse flies to the   areas
south of the Save river is either negleted or underestimated.
In our opinion the countries directly interested on the resolu
tion of this problem should endeavour to work together, inspite
of their political differences and the high costs involved.

b) Elimination of what we call the "Barué-Changara belt",
limited to the West by the so name "Zambeze belt", In Zimbabwe.
The control of this belt would contribute, not only to the cre
ation of an efficient buffer zone against the "Zambeze belt",
protecting it against eventual future spreads of the tsetse fly
from Mozambique, but also to the increase of the pasturage a-
reas in the Tete province, south of the Zambeze river, favou-
ring the natural development of the natural development of the
remainder cattle herds of the area.

These are the actions which, in oour opinion, should be given priority.

The struggle against the tsetse flies, necessarily carried
out in an integrated form, and taking into account the ecologi
cal conditions of the two belts under consideration, should con
sist of the following actions:

1 - Establishment of anti-tsetse fly barriers

One, south of the "Massangena belt" and of the "Gonarhezou
belt, and, two others to control the "Barue-Changara belt"from
the confing belts in the Mozambique territory.

The barriers to be built by mechanical means should be established in the areas with more xerophitic vegetation. They
should have a minimum width of 2 kilometers. Since such large
clearings would cause a considerable reduction of the photosyn
thesis, future reforestation should be considered using species
of rapid growth before pioneer vegetation begins to take hold.
A new rype of barrier would thus b formed which would be unsur
mountable to the tsetse fly.

2 - Establishment of wild life barriers

These would be, essetially, inter-territorial barriers, si milar to the ones built in the past along part of the West border of Mozambique, in the zone confining with South Africa and Zimbabwe.

These barriers which were built xpressly for sanitary reasons (prevention of the spread of the "foot and mouth disease" from certain wild life species to domestic animals) offer the advantage of interdicting normal transit of game across the various countries. It is therefore important to ensure the con - servation and improvement of the existing barriers, and build a new one along the Mozambique-Zimbabwe border, confining with the "Zambeze belt".

3 - Use of fly-traps

These should be placed along the periphery of the tsetse belts to be controlled. The traps should be of different types so that the most economical and efficient one can be found.

4 - Spraying operation with insecticides

The insecticides to be used, of which "endosulfan" has already given good results, should be sprayed from the air (by airplane or helicopter) or from the ground, by mechanical means..

Spraying from the air should tak place fundamentally in the "Barué-Changara belt", but only after it is isolated from the confining belts in Mozambique.

As we know that a similar operation is expected to be carried out in the "Zambeze belt", all the glossine communities of the "Barué-Changara belt" would be subjected to insecticide apraying from a combined action by the countries involved. As to the use of manual sprayers, we ony recommend it on he "Save belts", mainly on the breeding-places and on the "resting--grounds" of the main areas to be treated.

5 - Use of desinfestation chambers

These should be used to stop flies to enter in already free areas. They should be built in the traditional forms and placed on all the roads where traffic is normally slow due to the bad conditions of the pavements, thus allowing flies to be transported by passing vehicles. On tarred roads the use of the se chambers would not be compulsory, provided minimum speed li mits are enforced for motor vehicles. Cyclist, pedestriands and animal traction transportation would however have to comply with desinfestation rules.

Conclusion

We think if the methods exposed and considerations made are adequate and judiciously applied, it should be possible to obtain the recuperation of important areas in Mozambique, free from the harmful presence of the tsetse fly.

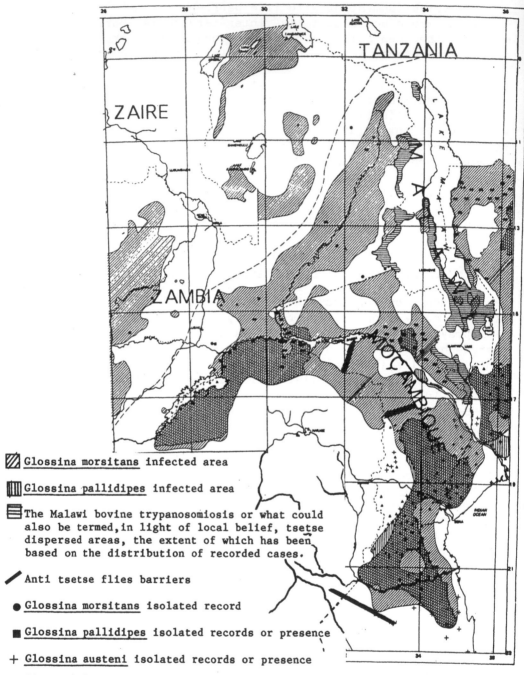

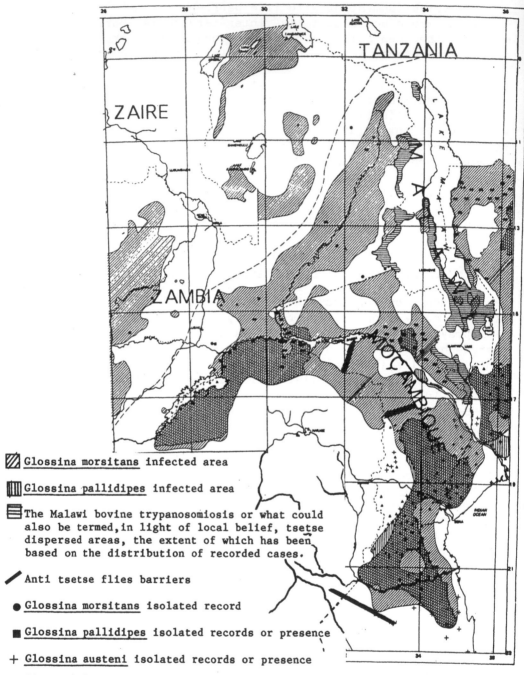 Glossina morsitans infected area

 Glossina pallidipes infected area

 The Malawi bovine trypanosomiosis or what could
also be termed, in light of local belief, tsetse
dispersed areas, the extent of which has been
based on the distribution of recorded cases.

 Anti tsetse flies barriers

● Glossina morsitans isolated record

■ Glossina pallidipes isolated records or presence

+ Glossina austeni isolated records or presence

▲ Glossina brevipalpis isolated records or presence

◆◆ Human trypanosomiosis or sleeping sickness endemic
foci and areas where cases occur sporadically

The approximate dividing line between the two sub
species of Glossina morsitans, G.m.centralis and
G.m.morsitans

REFERENCES

DIAS, J.A.Travassos Santos Dias:
    1962 - The status of the tsetse fly in Mozambique before
        1896. South Afr.J.Sc., 58(8):243-247.

HORNBY, H.E.:
    1947 - Report on a survey to ascertain if there be a dan-
        ger that the "Glossina morsitans" belt associated
        with the rio Save may spread South-West to the rio
        Limpopo and beyond. Lourenço Marques: 69

The sterile insect technique for tsetse eradication in Nigeria

W.Takken

Joint FAO/IAEA Division of Isotope and Radiation Applications of Atomic Energy for Food and Agricultural Development

SUMMARY

The results of a project whose objective was to eradicate Glossina pal-palis palpalis from a 1500 km^2 area in central Nigeria are reported. The project (BICOT) was conducted by a team of scientists including Nigerians and expatriates in Nigeria and by scientists from the tsetse unit of the International Atomic Energy Agency, Seibersdorf Laboratory, Austria. Within the project area, there was a total of 450 linear kilometers of gallery forest. Prior to the release of sterile insects, the wild tsetse population was reduced by more than 90 % with biconical traps. Sterile male flies for field release were produced by a mass-rearing facility in Vom, 200 km from the release area. Both in vivo and in vitro feeding tech-niques were used. Male flies destined for release were colour marked, irradiated at 12 k-Rad in a Co-60 source and allowed to feed on guinea pigs, before being packed and shipped to the release area. These were re-leased in the target area, approximately 150 sterile flies per week per linear kilometer of gallery forest. A sterile to wild male ratio of 3:1 was expected to lead to eradication, but when this proved ineffective, the ratio was increased. At the 10:1 ratio, eradication of G. p. palpalis was reached within 6-12 months. The impact of the sterile males on the wild population was studied by examination of the female reproductive organs, and confirmed by high levels of induced sterility. The eradication area was protected from reinvasion by the use of insecticide impregnated tar-gets, which had been placed alongside streams surrounding the area, at 50 m intervals over a length of 10 linear kilometers. At the time of writing (January 1986) 60 % of the area had been freed of G. p. palpalis using the SIT, while eradication in the remaining area was in progress.

1. INTRODUCTION

During the past 30 years, the tsetse fly has been eradicated from large areas of northern Nigeria by the use of persistent insecticides applied on those parts of the vegetation used by the fly as its dry season habitat, Only 10 % or less of the land had to be treated (1,2). Further South the fly occupied less well defined niches because of the different climate, and successful control necessitated a larger amount of insecticide per unit area because of the denser canopy of vegetation and higher rainfall. This not only increased the operational costs per unit area, but also caused concern about the fate of the environment exposed to larger amounts of hazardous pesticides.

In order to avoid the increased use of insecticides, the sterile insect technique (SIT) was selected as a potential method to be developed for the eradication of riverine tsetse and to be integrated with existing control methods. Since 1979, Nigeria and the International Atomic Energy Agency (IAEA) have jointly carried out a project (BICOT) for the development of the SIT against Glossina p. palpalis (Robineau-Desvoidy). The project has received considerable financial support from Italy, the Federal Republic of Germany, Belgium, Sweden and the United Kingdom.

The SIT is a method of insect control, whereby sterile flies are brought into an area to compete with wild flies for mating chances. If the sterile fly pressure is sufficiently high, sterility is introduced in the target population which will subsequently be reduced or become extinct. The method is more effective if the target population can be reduced before the sterile flies are brought into it.

One of the prerequisites of the SIT is the availability of large numbers of sterile flies, produced preferably under controlled conditions in a specialized factory. At the IAEA laboratories in Seibersdorf, Austria, research is undertaken to develop and improve massrearing techniques for tsetse flies. Also, studies are being made of the reproductive biology of tsetse flies to measure the sterility pressure in wild target populations during a SIT campaign. These activities have been directed for the past six years towards assisting BICOT, and much of the work presented in this paper was made possible because of the transfer of technology from Seibersdorf to Nigeria.

2. MATERIALS AND METHODS

2.1 Mass-rearing and sterile male production

The mass-rearing of G.p. palpalis was initiated in the Seibersdorf entomology laboratory of the IAEA, initially using guinea pigs as hosts, but later transferring the colony to an in vitro system on a 50/50% mixture of reconstituted freeze dried bovine/porcine blood. This technology was transferred to BICOT, where a mass-rearing plant had been built in Vom. For comparison purposes, two G.p. palpalis colonies were established blished in Vom, one in vivo (guinea pigs) and one in vitro (reconstituted freeze dried blood, 50% bovine/50% porcine). These colonies were to be developed to a size sufficiently large to produce the number of sterile males required by the eradication model. Fly rearing and related techniques have been described in detail (3).

Males to be released were radio sterilized in a Co-60 source at 12 k-Rad in air. Sterile males were colour marked and packed into 10 1 release containers. Before transportation to the field, 200 km away, sterile males were given an opportunity to feed on guinea pigs.

2.2 Study area

An area of 1500 km^2 was selected in central Nigeria, within the subhumid savanna belt 200 km from Vom. Being part of an agricultural development scheme, it consisted of a mixed farming/woodland area, intersected by 450 km of well-wooded rivers and streams. G.p. palpalis and G. tachinoïdes Westwood were present, the former species more abundant than the latter. At any time of the year, there were an estimated 40.000 head of cattle, mostly kept in a semi-nomadic manner. Wildlife was rare or absent.

Because of the pilot nature of the project, G. tachinoides was not considered for eradication and the species will not be discussed in this
paper.

2.3 Pre-eradication studies

Tsetse populations were measured from biconical trap catches, expressed as
the number of flies caught per trap per 24 h period. A minimum number of 5
traps per sampling area was used, usually placed at intervals of 250 m
along the streambed. At a nearby field station all fly catches were routinely examined and dissected for ovarian ageing.

The effect of continuous trapping with biconical traps to reduce the target tsetse population prior to the release of sterile males was studied
using different trap-intervals to obtain maximum reduction.

An experiment, designed to study the most efficient integration of tsetse
reduction by traps and sterile male release, was undertaken in four 5 km
long adjacent riverine forests. Each forest was to have a different period
of continuous trapping to reduce the tsetse population, followed by the
release of sterile males, initially at a ratio of 3 sterile males to one
wild male. Results of this study were used for the design of the eradication programme for the entire project area.

Tsetse barrieres were made of blue cotton screens, that had been impregnated with the insecticide deltamethrin (4,5). The impregnated screens
served as tsetse attractant targets. As barriers, screens were placed
along the streambed at 50-100 m intervals, suspended from metal rods.

2.4 Eradication plan

The results of the pre-eradication studies were used to develop a plan for
the eradication of G.p. palpalis. The plan was designed to reduce the
population as much as possible with traps, to be able to make efficient
use of sterile males on as large a scale as possible during the final
eradication phase. In areas where reinvasion might occur, insecticide
impregnated screens were placed over a length of 10 linear km of gallery
forest, 20 screens per km.

3. RESULTS

3.1 Mass-rearing and sterile male production

Results of the mass rearing of G.p. palpalis in BICOT are shown in figure
1. There was a large variability in colony performance of both the in vivo
and in vitro colonies. This was caused mainly by regular disease outbreaks
in the guinea pigs (in vivo) and by inadequate sanitary conditions (in
vitro). Following repeated adjustments these setbacks have been overcome.
The maintenance of a large back-up colony in Seibersdorf has helped to
sustain the BICOT colonies in times of low performance; e.g the sharp drop
in in vivo flies in 1985 was corrected quickly by the input of puparia
from Seibersdorf.

In late 1985, the BICOT colony of 100.000 female G.p. palpalis was producing 8000 sterile males weekly. Colony records were computerized to facili-

tate calculation of the number of male flies available for sterilization and field release.

3.2 Glossina p. palpalis natural distribution

G.p. palpalis was widespread throughout the project area. Its preferred habitat consisted of 25-200 m wide, dense evergreen gallery forest, which was common in all major tributaries of the Akuni and Feferuwa rivers (Figure 2). It was rarely found in vegetation whith an open understory, as was found in the main rivers. Fly densities varied from 0.10-19.30 fly per trap per day, depending on the locality and the season. High densities were usually associated with river crossings or watering points.

3.3 Tsetse reduction

Reduction of tsetse populations by continuous trapping was studied for periods varying from two weeks to 20 months. In all cases a sharp decline was observed within the first two weeks of trapping, reaching 90% or more after four weeks. Figure 3 shows the results of continuous trapping in two independent riverine forest. Although reductions of up to 97% were observed, continuous trapping did not result in eradication. A minimal trapping period of six weeks was found to give the maximum reduction desired before the introduction of sterile males.

3.4 Integration of tsetse reduction and sterile male release

The forest patches Kirayi, Tsakuwa, Maiakuya and Maisamari (Figure 2) were each subjected to different periods of trapping, followed by sterile male release. Sterile males were released at weekly intervals from pre-determined points, approximately 2 km apart; 150-300 flies were released per point, depending on the required sterile male pressure. Sterile males did not behave differently from wild males, moving rapidly through the gallery forest and concentrating at the same points as the wild flies. Two weeks after release, 50% of the released flies were still alive, whereafter there was a rapid decline in recaptures. Following six weeks of continuous trapping, a sterile to wild ratio of 3:1 could be established. However, this ratio proved too low to provide sufficient pressure on the wild population, as was observed from the low incidence of induced sterility in wild female flies. When the ratio was increased to 10:1, the SIT became effective. Using this ratio, a period of 6-12 months was required to eradicate the target population from these four reverine forests, which have since then remained fly free (Figure 4).

3.5 Tsetse barriers

Insecticide impregnated screen barriers proved effective in reducing resident tsetse populations to low levels (90% reduction or more), but also acted as interceptors of flies moving into the barrier from elsewhere. The barriers effectively kept flies out of the Akuni basement, thus protecting the four experimental eradication areas (Figure 2).

3.6 Project status in January 1986

The Akuni basement enclosed by the project boundary has been free of G.p. palpalis since early 1985. The area has been surveyed monthly since then (Figure 4). In the Achiba/Dungu/Akwamino complex, high levels of sterility have been recorded since March 1985. By August 1985 the population had collapsed and in January 1986 flies were no longer found (Figure 5). Tsetse populations in the remaining project area have been exposed to continuous trapping since February 1984 until weekly sterile male release began in January 1986.

4. DISCUSSION

BICOT has demonstrated that G.p. palpalis can be eradicated using the SIT. Two thirds of the 1500 km^2 target area is now tsetse free, and eradication is expected in the remaining areas during 1986. Other riverins tsetse, G.p. gambiensis and g. tachinoides, were eradicated from an area in Burkina Faso using the same technique (6).

It is encouraging that the mass-rearing technology is no longer a bottleneck for the application of the SIT against tsetse. Recent developments in the use of locally collected blood, as well as computerized colony record administration, have further reduced the costs of this aspect of the SIT. The presence of a back-up colony has nevertheless proven to be invaluable in the case of BICOT.

The reduction of the target population by the use of continuous trapping is a simple and low-cost method of insect control, avoiding the use of aerial or ground spraying of insecticides. This method can be made more effective by the impregnation of traps with insecticides (7) or by replacing traps with insecticide impregnated screens.

Sterile to wild male ratios had to be higher than Knipling's original model (8), which predicted eradication at the 3:1 ratio. This was also found in Burkina Faso (6), and it appears that in a SIT programme a 10:1 ratio is required to eradicate riverine tsetse in West Africa. The entire BICOT area could not be flooded with sterile males simultaneously because of the limited colony size. This, however, may not have been disadvantageous, because it was found that individual, isolated riverine complexes could be freed of tsetse using the SIT in a sequential manner. Insecticide impregnated screen barriers, which can be operated by unskilled personnel or even the local population, seem effective as a temporary or permanent measure to isolate eradicated areas from tsetse infested areas.

5. ACKNOWLEDGEMENTS

I thank Dr. D.A. Lindquist and Dr. A.M.V. van der Vloedt for useful advice on the preparation of the manuscript.

REFERENCES

1. DAVIES, H. (1971). Further eradication of tsetse in the Chad and Gongola river systems in North Eastern Nigeria. J. appl. Ecol. 8, 563-578.
2. SPIELBERGER, U., NA'ISA, B.K. and ABDURRAHIM, U. (1977). Tsetse (Diptera:Glossinidae) eradication by aerial (helicopter) spraying of persistent insecticides in Nigeria. Bull. ent. Res. 67, 589-598.

3. TENABE, S.O. et al. (in press). Status of tsetse control by the sterile insect technique in Nigeria.
    II. Advances in the mass rearing of Glossina palpalis palpalis in International Scientific Council for Trypanosomiasis Research and Control, 18th meeting, Harare.

4. GOUTEUX, J.P. et al. (1982). L'utilisation des écrans dans la lutte anti-tsetse en zone forestière. Tropenmed. Parasit. 33, 163–168.

5. OLADUNMADE, M.A. et al. (1985), Studies on insecticide impregnated targets for the control of riverine Glossina spp. (Diptera: Glossinidae) in the sub-humid savanna zone of Nigeria. Bull ent. Res. 75, 275–281.

6. CUISANCE, D. et al. (in press). La campagne de lutte intégrée contre les glossines dans la zone pastorale d'accueil de Siderodougou (Burkina Faso) in International Scientific Council for Trypanosomiasis Research and Control, 18th meeting Harare.

7. LAVEISSIERE, C. and COURET, C. (1980). Traps impregnated with insecticide for the control of riverine tsetse flies. Trans. R. Soc. Trop. Med. Hyg. 74:2, 264–265.

8. KNIPLING, E.F. (1963). Potential role of the sterility principle for tsetse fly eradication. Who Vector Control 27, 17 pp.

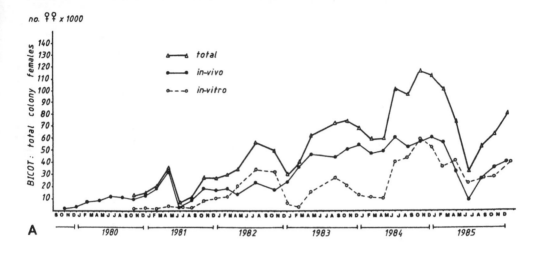

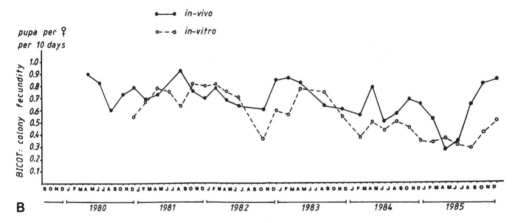

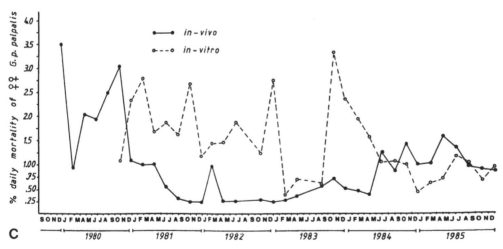

FIGURE 1 – Performance of the BICOT _in vivo_ and _in vitro_ colonies of
G. p.palpalis. A- colony size; B- fecundity; C- mortality.

89

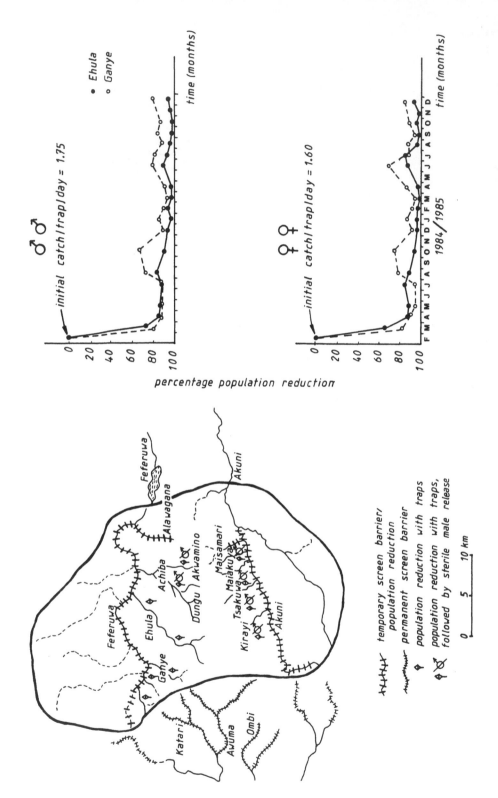

FIGURE 3 – Reduction of G. p. palpalis in Ehula and Ganye due to removal trapping.

FIGURE 2 – Riverine systems of the BICOT study area, and method of tsetse control/eradication applied.

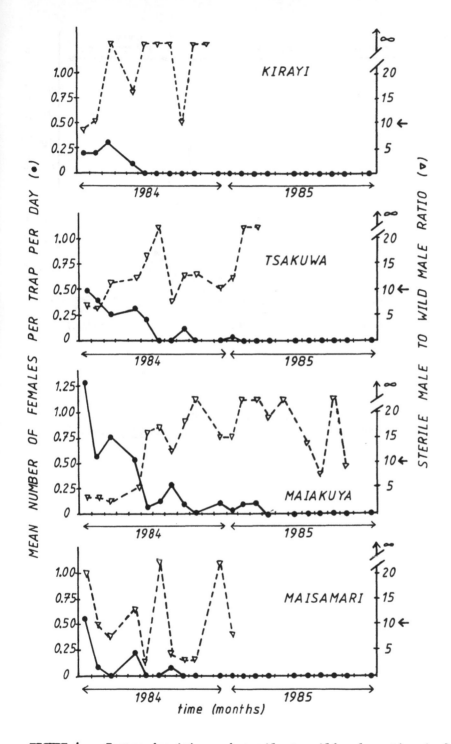

FIGURE 4 – Tsetse densities and sterile to wild male ratios in four tributaries of the Akuni river. The arrows indicate the critical value of sterile male pressure.

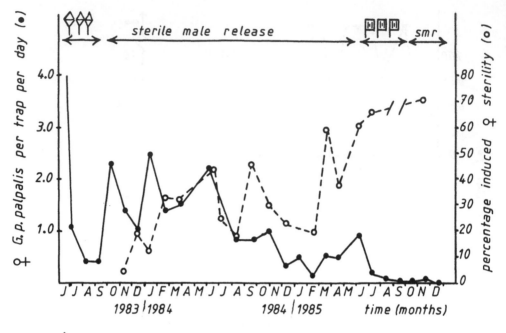

FIGURE 5 – Tsetse density in the Achiba/Dungu/Akwamino riverine complex and level of sterility induced in the wild female G. p.palpalis population.

Biotechniques and other methods of tsetse fly control

D.A.Dame
Agricultural Research Service, United States Department of Agriculture, Gainesville, Fla.

Summary

The selection of combinations of control methods to incorporate into an overall tsetse management strategy must involve and in-depth assessment of the strengths and weaknesses of each of the individual components as well as the composite package. Requirements common to all control techniques, e.g., isolation from reinfestation, and some limitations of specific methods are presented. The value of dedicated research support units and computerized models to cope with the complexity of long term control programs is emphasized.

Introduction

The significance of trypanosomiasis in tropical Africa has been documented adequately in this symposium and elsewhere. Because of their role as vectors of African trypanosomiasis, several species of Glossina have been targeted for elimination from locations where disease transmission limits livestock development or human health. In many of these locations more than one species is incriminated and control programs must be directed at infestations in a variety of habitats. Thus, control practitioners need to address the question of integrated management systems.

Limiting factors

Allsopp directed attention to this issue in his review of insecticidal control of tsetse flies(1); he is not alone in advocating that insecticidal methods be supplemented with nonchemical approaches when appropriate. The proven control methods that are currently available include aerial and ground applications of aerosols or residual deposits, traps impregnated with insecticides and supplemented with effective volatile attractants, the sterile insect technique, habitat modification, and host animal suppression. Not all of these

93

methods are currently being practiced. Examples of novel methods that might become available for future use against the major vectors include more sophisticated genetic approaches, expanded use of attractants, perhaps in combination with synthetic pheromones and chemosterilants, and immunization against the trypanosomes.

Yet I believe that the currently existing technology provides enough complementary methodology to conduct integrated tsetse management in numerous locations currently considered to be high priority. However,to realistically confront the problems involved in managing tsetse populations, it is important to think in terms of large scale operations which include natural or artificial barriers to reinfestation. This implies that long term considerations must be involved in most cases because of the cost and staff commitments required of large scale programs. An excellent example of a long term, large scale commitment is the screwworm (Cochliomyia hominivorax) eradication program in North America. Following a successful research program in the early 1950's, this veterinary pest was eliminated from the island of Curacao and by 1959 from the state of Florida. The eradication costs in Florida were ca. $8,000,000, a favorable amount when compared to the estimated annual losses to this pest of ca. $20,000,000 in Florida alone. The program moved to the southwestern United States, where combined governmental and rancher funding was used to support the release program and to maintain a lengthy barrier over 200 miles wide by continued releases between the infested and the treated areas. When this barrier proved ineffective, the program was extended southward and the fly now has been pushed to lower Mexico, with plans to move the program all the way to the southern border of Panama. Approximate costs of this program were recently estimated at $500,000,000 through 1984, with an estimated savings to the industry of $5,000,000,000. The goals of this program will have been accomplished when the fly has been eliminated from Panama and an effective sterile male release barrier has been established there. This is an example of long term commitment and large area concept in a successful program that has not been without controversy and budgetary constraints related to mission accomplishment.

This example is cited not to suggest that similar efforts be devoted to tsetse control, but rather to illustrate the need to be committed to long term strategies, if they are required. The example also reinforces the need to deal effectively with population isolation, a key element in the management of tsetse populations regardless of the control methodology.

When assessing the acceptability of specific control methods it is not unknown for reviewers to ascribe certain requirements to one method, but to lose sight of the fact that the same requirements may apply to other methods as well. The issue of isolation is a good example. I have often been reminded that isolation is a limiting factor and an essential requirement for successful completion of a sterile insect program. It is true that this is a major factor to be considered when dealing with management of population reproduction. However, it is also an essential factor when dealing with each of the other control methods. Thus, if a proposal for utilization of the sterile male technique were rebutted because of inadequate isolation, then a proposal for the use of another method would be subject to the same criticism.

There are a number of other requirements or factors that are common to all control approaches. For example, the environmental impact of a proposed technique should be as low as is possible and be at an acceptable level. The institutional infrastructure should have sufficient capability and depth for a long term control effort, and cooperation of the neighboring countries should be assured if isolation cannot be achieved within the national boundaries. Where multiple species are targeted for control, the method(s) proposed should be adequate to accomplish this objective. When possible, subsequent usage of the cleared land should be carefully planned; it should be recognized, however, that when large areas must be involved in order to cope with isolation issues, it may not be possible to satisfy all constituents in this regard. And other factors that restrict subsequent land usage should be carefully assessed, e.g., limitations due to tick-borne disease, soil fertility, social customs, etc. Also, costs and support logistics should be feasible for the long term.

Each specific control methodology has limitations or specialized requirements that must be considered in the decision-making process that precedes selection of the strategies to be integrated. For example, aerial and ground application of residual deposits of insecticides should be assessed in terms of their potential impact on nontarget organisms and the environment. Aerial application of insecticide aerosols should be assessed in terms of the capability of dispersing the proper droplet size spectrum for vegetative penetration in the target terrain at the appropriate season. The capability of producing and distributing viable and competitive sterile insects in numbers adequate for the task should be assessed when considering the sterile insect technique; the additional costs of dealing with habitats with multiple species must be also considered. If traps are to be used, the safety of the insecticides or chemosterilants employed should be assessed in relation to human and other nontarget risk.

Having devoted a considerable amount of personal attention to the sterile insect technique, I would like to address two issues that seem to me to have raised an unnecessary level of controversy. First, there is the issue of multiple mating behavior, which is erroneously considered

95

by many to be an unescapable factor that limits the usefulness of the sterile insect technique. It is true that if the sperm of irradiated or chemosterilized insects are not competitive with sperm from untreated males in the indigenous insect population, multiple mating behavior on the part of the females could severely reduce the impact of sterile mating events. However, when the sperm of the treated males are competitive their effect is not diluted by multiple mating behavior. This is due to the fact that the incidence of sterile mating is directly proportional to the ratio of effective sterile males to indigenous males. Thus, if there is one effective sterile male for each indigenous male 50% of the mating achieved will be sterile; an indigenous female that has already mated with a sterile male has a 50% chance of mating again with a sterile male in each subsequent mating. To date there is no evidence to indicate that released tsetse males, irradiated or chemosterilized at the proper exposure levels, produce sperm that are not competitive. In fact, the existing evidence indicates that the sperm are competitive and that, where ratios of one sterile male to one fertile male have been established in the field the expected reduction of reproduction has been achieved (8).

The second issue for discussion relates to the impact of released sterile insects on trypanosome transmission. It has been amply demonstrated the sterile tsetse flies can transmit trypanosomes and that they thus pose a risk to their hosts (2,6). However, the risk will be proportional to their abundance in the release zone. No data have been generated that suggest that released flies are more likely to transmit trypanosomes than their natural counterparts. Program managers have the capability of using strategies that reduce indigenous tsetse populations far below their normal densities prior to the release of sterile insects to cope with those reduced populations. The net result of such a strategy could be to experience combined sterile and indigenous fly densities far below those which existed prior to initiation of the program. An immediate reduction in transmission of trypanosomiasis should occur, followed by the complete cessation of transmission at the end of the control program. In fact, this may be the most cost-efficient strategy because it takes advantage of the high level of cost effectiveness of insecticides when the population density is high and of sterile insects when the population density is low. Furthermore, since it has been demonstrated in the field that ratios of sterile to fertile males of 1:1 do bring about the expected 50% reduction in the reproduction of the indigenous tsetse populations, it should not be necessary to release overwhelming ratios of sterile tsetse in areas where transmission occurs. Thus, it seems that the potential for trypanosome transmission by released insects is an issue that could be dealt with at the program level to ensure that there would be a progressive reduction in transmission.

Dedicated Research Support

An essential component of long term control programs is a realistic commitment to maintain a research capability associated with, but separate from, the operational staff. To again use the screwworm program as an example, a dedicated research team has been associated with the program from its inception, first in Kerrville, Texas, then in Mission, Texas, and currently in Tuxtla Gutierrez, Mexico where the production plant is located. This team conducts research related to program needs that would be difficult for the operational staff to fulfill because of their routine commitments to the program. For example, when a new screwworm strain is required, because the releases have been moved to new locations or there is doubt concerning the viability of the existing strain, the research team field-collects the material, initiates colonization, and assesses the strain in the laboratory. Then, in collaboration with the operational staff, pilot scale production runs and field evaluations are conducted prior to making the final decision on strain replacement. Thus, this team is responsible for collection of data that lead to program improvement, and provides information and advice for coping with the changing needs of the program. The existence of this dedicated unit enhances opportunities for involvement in new creative research and provides essential research support, without which the program would suffer in terms of both costs and mission accomplishment.

Simulation Models and Computer Assisted Interpretations

Population simulation through the use of computer models provides a powerful tool with which managers of control programs can plan strategy, follow progress and revise action plans rapidly. A number of mathematical models have been created to describe the response of tsetse populations to environmental or control stresses [7]. Also, a number of life-cycle based computer models have been developed for simulating natural populations of insects of veterinary or medical importance [4,5]. Some of these models, e.g., the lone star tick (Amblyomma americanum) model, are now available for worldwide distribution and use on microcomputers[3]. These comprehensive programs can have the capacity to accept informatiion on life stage, survival, temperature, rainfall, photoperiod, and diapause, to name a few useful parameters. Use of the models can provide important information on which biotic and abiotic factors influence population dynamics, and at the same time can indicate which parameters do not merit in-depth investigation. They can help to identify information gaps as well as allow managers to predict population responses to pest management practices. Properly used, they can reveal which pest management strategies have potential and which are not worth the investment of time and resources for study. The development of user-friendly models for simulation of tsetse fly populations is imperative if we are to make knowledgeable decisions on management strategies, to have a clear understanding of the impact of integrated control methods on population density and trypanosome transmission, and to determine program end points.

Perhaps one of the most difficult tasks in deciding which control methods to include in a management strategy is to identify the candidate methods that are sufficiently proven to be acceptable for the program at hand. This may be more difficult with tsetse than with other target species because of the lack of isolation in most experimental scale trials. For example, studies on the efficacy of the sterile insect technique or of aerial insecticide application require sizeable experimental plots and are very costly to conduct. Yet, the plots are subject to reinfestation from outside. In recognition of the significance of this factor, researchers collect data during the course of their experiments to help interpret the impact of this migration on the results. With this data, it is then possible to estimate the true effect of the experimental treatments. Usually the conclusions can be considered to be statistically definitive, but they may still be subject to some low probability of error in interpretation. Unfortunately, often the only way to confirm the conclusion is to test it operationally, because it is too costly to conduct further research – which would again be subject to the same limitations. In retrospect, this aspect serves to reconfirm the significance of the joint concept of large scale and isolation as an important feature in integrated management of tsetse populations.

Summary

In these discussions I have attempted to focus on the fact that there are limitations and restrictions with each of the available control methods. Some of these limitations are common to all of the techniques, and some apply only to a specific approach. The selection of combinations of control methods to incorporate into an overall management strategy must involve an in-depth assessment of the strengths and weaknesses of each of the individual components as well as the composite package. Finally, experience with large scale and long term programs has confirmed the value of dedicated research support units and the need for computerized models and support systems to cope with the complexity of the task.

REFERENCES

1. ALLSOPP, R. (1984). Control of tsetse flies (Diptera: Glossinidae) using insecticides: a review and future prospects. Bull. Ent. Res. 74: 1-23

2. DAME, D.A. and MACKENZIE, P.K.I. (1968). Transmission of Trypanosoma congolense by chemosterilized male Glossina morsitans. Annals Trop. Med. Parasit. 62: 372-74

3. HAILE, D.G. and MOUNT, G.A. (in manuscript). Computer simulation of population dynamics of Amblyomma americanum (L.) (Acari: Ixodidae)

4. HAILE, D.G. and WEIDHAAS, D.E. (1977). Computer simulation of mosquito populations (Anopheles albimanus) for comparing the effectiveness of control technologies. J. Med.Ent. 13:553-67
5. MILLER, J.A. (ed.) (1986). Modeling and simulation: tools for management of veterinary pests. U.S. Department of Agriculture, Agricultural Research Service, ARS-46, 54 p.
6. MOLOO, S.K. and KUTUZA, S.B. (1984). Vectorial capacity of gamma-irradiated sterile male Glossina morsitans centralis, G. austeni, and G. techinoides for pathogenic Trypanosoma species. Insect Science and its Applns. 5: 411-14
7. ROGERS, D.J. and RANDOLPH, S.E. (1985). Population ecology of tsetse. Ann. Rev. Entomol. 30: 197-216
8. WILLIAMSON, D.L., DAME, D.A., GATES, D.B., COBB, P.E., BAKULI, B. and WARNER, P.V. (1983) Integration of insect sterility and insecticides for control of Glossina morsitans morsitans Westwood (Diptera: Glossinidae) in Tanzania. V. The impact of sequential releases of sterilized tsetse flies. Bull. Entomol. Res. 73: 391-404

Session 3
Present and future programmes of international and national organizations

Chairman: J.Mulder

Present and future programme of GTZ on tsetse and trypanosomiasis control

F.W.Brückle

German Agency for Technical Cooperation, Division of Veterinary Medicine, Eschborn, FR Germany

Africa's rapid population increase and the growing demand for a higher standard of living are leading to progressive meat shortages. Parallel to this development is an automatic rise in meat prices, causing a further reduction in the protein supply of the poorer groups of the population.

Even if a considerable increase in animal productivity and a radical improvement in marketing could be achieved in the near future, the meat supply at the present level of the animal population would hardly satisfy the rising demand.

Conventional farm animals are the main meat producers. Other sources are economically unimportant. The exploitation of game population for venison is a positive argument for the preservation of game animals, but it would be unrealistic to consider it primarily as a large scale contribution to the general meat supply.

Just these traditional farm animals, however, are being continually decimated by a host of endemic and epidemic diseases. Among them are rinderpest, pleuropneumonia, tickborne haemoparasitoses, endoparasitoses, and especially trypanosomiasis. In trypanosomiasis there are more correlations and interactions between predisposing environmental factors than is the case in other epizootic and endemic diseases. Close relationships also exist between endemic trypanosomiasis and land usage for animal production, with the danger of erosion in intensively grazed and trampled areas.

The much-cited figures that impressively reflect the importance of trypanosomiasis in Africa are certainly known to you: 2/3 of the surface of Subsaharan Africa is afflicted with trypanosomiasis and therefore hardly if at all usable for the production of domestic livestock. It is estimated that the present annual meat production of 1.5 million tons could be doubled if this obstacle were removed. The presence of the tse-tse fly prohibits the use of the fertile hot-humid pasture zones especially.

The result – putting it in a simplified way – is that Africa's vital cattle populations are restricted to the border areas between quasi vegetation-free arid zones and those regions made inaccessible by tse-tse infestation. Thus herds are concentrated in areas that are free of tse-tse flies by virtue of insufficient shade and lack of humidity. Such terrains barely support an increase of cattle. Progressive soil erosion and the intrusion of poorly nourished herds into more humid regions is the unavoidable result. Because the tse-tse fly pressure prevents overall distribution of livestock in potential pasture zones, stocking rates are raised in these border areas and the burden on the vegetation increases. The result is a gradual progression of the arid regions.

Only opening up of pastures on a large-scale basis can lead to a better distribution of cattle, reduce the burden on soil and vegetation, and halt the progressive erosion. It is therefore suggested here that it is not the tse-tse control, but rather the omission of tse-tse control that promotes erosion. This conviction has not changed over the past 15 years - the period in which GTZ has actively been engaged in tse-tse control. It might appear to others that GTZ has reduced its engagement in the combat of trypanosomiasis and tse-tse flies in recent years. This is certainly merely due to the fact that there are presently no large-scale, spectacular eradication campaigns being carried out. Projects using helicopters over those vegetation zones which allowed selective spraying with insecticides were a success in each case. They began in Niger in 1965-1973 and were continued in Nigeria from 1971-1978, where 12,000 km² were freed of tse-tse. The most recent campaign was in Cameroun, where, beginning in 1976, 16,000 km² in the Adamaoua highlands were opened up for cattle production in the course of an 8-year campaign. As is known, follow-up measures were unfortunately neglected in Nigeria, allowing part of the sprayed area to revert to glossina.

The campaign in Cameroun was accompanied by detailed and thorough investigations on the possible toxic effects of the insecticides. The Institute of Biogeography at the University of Saarbrücken was commissioned to undertake the observations. They concluded that in the short run, an anticipated and perceptible reduction in certain species did occur. Most of these groups, however, had recovered after the following rainy season. Some individual animal species, such as the bat, were not spotted until 2 years later. The shrew first reappeared after 3 years. Four years after spraying follow-up investigations showed that the re-immigration of eradicated species had brought back stabilization. The riverine-forest ecosystem was thus temporarily depleted, but not irreversibly damaged, thanks to the potential for regeneration and re-invasion from the surrounding area.

The use of insecticides certainly entails risks. We believe, however, that we can minimize these risks and advocate such campaigns by following exactly the advice of competent ecologists as to the type of insecticide and the method of application.

As far as I know, no model for a cost-benefit analysis regarding tse-tse control by helicopter has been developed so far. I am convinced, however, that spray campaigns can be economically justified provided that the land is usable and is appropriately used, and that follow-up measures are carried out properly to insure against the re-invasion of tse-tse. The present problem with large-scale spray actions thus seems to lie neither in the sphere of ecology nor in that of profitability, but first and foremost in the current high demand for capital, which becomes continually more difficult to mobilize.

Since 1974, while GTZ was still involved in the spray campaigns in Nigeria, and before the project in Cameroun began, possibilities of controlling or eradicating tse-tse fly by using the sterile-insect technique (SIT) have been examined. A joint research project with France was initiated in Bobo Dioulasso. After prestudies and subsequent larger-scale field trials had shown clearly positive results, tse-tse eradication was carried out in a 3,000 km² area around Sideradougou in Burkina Faso. At the conclusion of this experiment it was proved that the SIT led to the desired result, and that the tse-tse fly had vanished from the area. Of course the imperative attention to long-term protection against invasion is just as important here as with the use of insecticides. In general this factor produces more worries than the control itself. The breakdown of protective measures that are no longer under our influence very quickly wipes out previous successes and makes the investment appear wasted. This has a long-lasting effect on political and financial decision-makers. One project subsequently ruined in this manner causes more damage in the field of judgement of political and economic decision-makers than two or three permanently successful projects can rectify.

We will not go into the SIT any further. This was the subject of Dr. Politzar's report. I would merely like to add that GTZ is currently setting up a comprehensive study on the economics of the SIT. The study will be finished by about the middle of 1987.

For approximately 4 years GTZ has been working on tse-tse control with traps and screens in two countries: in the already mentioned German-French project in Bobo Dioulasso and in the northern Ivory Coast. The idea in itself is not new. I already used traps 30 years ago in the then Belgian Congo, although with somewhat discouraging results. In contrast, the present trapping technique has been developed to such an extent as to justify great hopes. The recent results observed in the Ivory Coast and Burkina Faso make us optimistic enough to consider the traps and screens method the means of choice in many cases. This will be emphasized even more once the attractants are available for large scale use in widespread areas. The European Community is financing a research project on this issue and France, Great Britain and the Federal Republic of Germany are participating in it.

Two further aspects of the trapping technique appear to me to be of great importance:
- the possibility of small-scale activity in the vicinity of a single ranch or village, at water holes or other selected targets; this represents a substantial advantage over insecticides sprayed from the air, and
- the possibility of an active material participation on the part of the people who benefit. The degree of this participation, which would ideally be complete responsibility, is certainly dependent on socio-economic factors. Studies in the Ivory Coast were initially not very encouraging, however. Although both the Peulh and the Senufo quickly recognized the advantage of the traps, their participation in the action fell short, due to a declining sense of community spirit and increasing individuality and egoism.

Nevertheless I would like to warn against excessive enthusiasm, despite the very promising results. The full extent of the limitations in trapping is perhaps still to be seen. Trapping will certainly not be feasible in areas where insufficient accessibility for the installation, checking, repair and replacement of traps impedes or prevents their use.

However, GTZ does not wish to rely on the control of the tse-tse fly alone. Diversification of engagement as far as possible was intentional. For this reason GTZ is deeply involved in a German-French research project in Bobo Dioulasso concerning the immunology of Trypanosomiasis.

A scientific exchange of ideas in this special area exists with ILRAD. We hope that this cooperation will become closer and more pragmatic.

GTZ has demonstrated at Bobo Dioulasso the extent of its interest in this approach with a 7-year programme which will be extended in 1986 for another 3 years. To date 10 million German Marks have been made available for this research and it is expected that this will be increased by an additional 3 million. Furthermore GTZ personnel have been working on the improvement of trypanosomiasis chemotherapy in Berlin and Kenya since 1978. In Berlin a screening lab is available to all interested synthesizing laboratories of the pharmaceutical industry and public research institutes that are willing to cooperate. Here new and promising molecules can be tested for possible trypanocidal characteristics. Nine companies sent in a total of 2094 substances for testing during 1984.

In a project in Kabete, Kenya, the potential new trypanocides undergo further testing regarding their toxicity, trypanocidal properties, optimal form of application, and the development of resistance and cross-resistances.

In conclusion I wish to emphasize that GTZ's growing involvement in the control of trypanosomiasis and tse-tse fly during the last 10 years was not a passing whim. If anything, the importance of trypanosomiasis has grown as a result of increasing economic considerations.

In the course of the past 10 years tse-tse control has entered profoundly into the discussion of environmental problems. Arguments were not always objectively presented by professional opinion-makers. Thus the necessary development of an awareness of the difficulties and dangers accompanying tse-tse control was unfortunately hindered. In some cases, an almost sectarian journalism has led to a general rejection of tse-tse control.

GTZ will continue its efforts to combat trypanosomiasis as long as the financial means are available to do so. Certainly it is planned to continue the work by following several paths simultaneously. The best programmes for combating tse-tse flies will always be tailored to the particular circumstances. No method is yet known to us that would be equally suitable for the varying conditions of landscape, climate, economics and ethnic groups.

At this point I would like to make a plea to you. Top-level science, tenacious field work and enormous financial sums have been invested in research and development in this area by the participating nations. We owe it to them to exploit every possibility for cooperation, whether national or international, scientific or practical, financial or human, in order to achieve the common goal more rapidly.

LIST OF PREVIOUS ACTIVITIES IN THE FIELD OF TSE-TSE AND TRYPANOSOMIASIS CONTROL

Name of project and country Project Numbers	Type of project and duration	German funds involved in DM	Description of activities
Tse-tse program Phase I Ivory Coast 77.2252.3	Survey and ecology 1978-1987	5,931,731.-	Biological survey on the occurrence of glossina spp. Detailed tracing out of the distribution as a basis for ensuing control.
Tse-tse Program Phase I Ghana 77.2214.3	Survey and ecology 1979-1982	3,340,077.-	as above
Land utilization Cameroun 79.2143.0	Land utilization 1979-1980	668,392.-	Well conducted utilization of areas free from tse-tse flies.
Control of tse-tse flies in the Adamaoua highlands Cameroun 74.2180.3	Control 1976-1985	8,905,812.-	Continuation of Nigerian tse-tse control in adjacent areas in Cameroun. Preliminary work of German experts in collaboration with the World Bank started in 1975. It involved entomological and biological mapping of project area which covered about 16,000 km².

Project	Period	Amount	Financing / Description
Ecological control unit Cameroun 78.2525.0	Ecology 1979–1980	475,699.–	**Financing** Phase I, 1976 – 1979 GTZ – technical & scientific management – helicopter – insecticides World Bank – local costs Cameroun Phase II, 1980 – 1982 GTZ – technical and scientific management KfW (German credit) – helicopter World Bank – insecticides Cameroun – local costs Ecological survey on environment to determine effects of insecticides used in tse-tse control.
Trial of chemo-therapeutics against trypano-somiasis Supra-regional Kenya 75.2041.4	Research 1978–1987 1985–1987	7,326,277.– 2,650,000.– 9,976,277.–	Pharmaco-kinetic tests of new trypanocides. Dosage trials, application and accumulation in tissue. Scientific collaboration with the University of Erlangen.

Project	Type / Years	Amount	Description
Team of agricultural advisors Kenya 63.2046.9	Research 1965–1979		Research on trypanosomiasis was carried out among other animal diseases at the research institute in Kabete.
College and Research Institute for cattle husbandry Mokwa, Nigeria FE 406	Ecology and control 1962–1974		The project incorporated its own tse-tse department which carried out surveys on the ecology of tse-tse flies. Later control of tse-tse flies was carried out by the use of chemical and physical means within and outside the farm.
Tse-tse control Kaduna, Nigeria 73.2067.4	Practical control 1971–1978	8,867,000.-	Tse-tse control with helicopters to create new pastures. Research on and trials with various methods and insecticides in order to lower costs of tse-tse control.
Consultant plus equipment Control of epizootic diseases Niger FE 7312	Practical control 1965–1973	1,936,515.-	Control of tse-tse flies in the Niger river valley & its affluents. Preliminary research involved the examination of blood smears to determine the rate of infection in bovines, but also the collection of data on glossina populations in the south of the country. Thereafter treatment of trypanosomiasis and spraying in tse-tse areas with insecticides were carried out. One selected area (85 km of riverine forest) was completely cleared of tse-tse flies by meanns of helicopter and ground spraying teams. Bush-clearing alongside prevented reinfestation. Up to now this area is free of tse-tse.

Training Center for tse-tse Supra-regional Burkina Faso 75.2069.5	Training 1976–1984	7,690,927.–	With the creation of a supra-regional training center requirements for practical tse-tse control were to be met. Preliminary ecological studies & control itself required a one-to-two year training of veterinary assistants in the tse-tse region and in an autonomous training center geared to this kind of training. The need for such a supra-regional training center is demonstrated by the high number of applicants. German-French joint project.
Tse-tse Survey Burkina Faso 77.2178.0	Survey and ecology 1978–1979	1,732,537.–	Biological survey on occurrence of glossina spp. & detailed tracing-out of distribution as a basis for following control measures.
Immunology of trypanosomiasis Burkina Faso 77.2227.5	Research 1978–1988	14,005,276.–	Research on trypano-tolerance of certain cattle breeds which could create a basis for eventual future production of a vaccine against trypanosomiasis. This is a joint venture of GTZ and IEMVT (Maison Alfort). The German experts are integrated professionally into the working program of the IEMVT. Formally they remain a substantial part of the GTZ as written down in the bilateral agreement. Collaboration with the Max-Planck-Institute for Molecular Immune Biology, Freiburg.

Biological tse-tse control in Bobo Dioulasso Burkina Faso 73.2069.0	Research 1974–1987	9,166,873.–	Biological tse-tse control involving the rearing, sterilization and ensuing release of males. Control measures in an extended area of 3,000 km². Joint venture with IEMVT, Paris.
Central Veterinary Laboratory Tanzania 62.2135.2	Research 1964–1976		– Investigations were carried out to determine pathological and histological changes of trypanosomes in cattle. These investigations involved protozoological aspects too. – Cultures of trypanosoma strains in small laboratory animals and domestic animals for haematological and clinical trials. – Therapeutical and prophylactical trials.
Research Center for trypano-tolerance Togo 78.2157.2	Research 1978–1982		Started in 1964, the project acquired a new concept in 1977. The former "Centre Experimental d'Elevage" was transformed into "Centre d'Elevage et de Recherche sur la Trypanosomiase et la Trypanotolérance". Aim: investigations into trypano-tolerance of autochthonous cattle breeds and their cross-breeds. The research of tolerance mechanisms is followed by transformation of acquired knowledge on husbandry and rearing of trypano-tolerant cattle in the humid West African Savannah (rainfall 1200 – 1600 mm p.a.!).

Examination of Trypano-tolerant effects of chemical agents Supra-regional BGA-Berlin 79.2247.9	Research 1980-1987	2,862,405.-	First step in the development of new drugs against trypanosomiasis.
Comparative ecological study on different methods of tse-tse control. Supra-regional Ivory Coast 80.2068.7	Ecology 1980-1986	3,204,081.-	Preservation of the ecological balance in the control of tse-tse flies.
Animal Health Research Center Entebbe Uganda 70.2066.2	Research 1970-1977	1,409,000.-	In collaboration with the Tsetse Control Division: Investigation into the repeated occurrence of trypanostoma-infections in certain areas free of tse-tse flies by means of serological methods and various colour techniques.
Comparative & ecological study on different methods of tse-tse control Supra-regional 80.2068.7	Ecology 1981-1987	3,204,081.-	Accompanying study on the use and trial of different insecticides.

The ICIPE's research programme on tsetse and trypanosomiasis

J.Mutuku Mutinga

International Centre of Insect Physiology and Ecology (ICIPE), Nairobi, Kenya

SUMMARY

The objectives of ICIPE's research on tsetse flies and Trypanosomiasis are development of integrated, environmentally safe control strategies based on use of traps, potent attractants, tsetse pathological agents. The approach to these techniques are discussed in detail.

Relatively little impact has been made in the control of Trypanosomiasis despite many years of research and many attempts to control the disease or the tsetse flies. The disease and indeed the flies remain a great draw back to development in many areas of Sub-Saharan Africa. It is estimated for instance, that out of 570.000 km^2 land area of Kenya, nearly 138.000 km^2 is tsetse infested. It is quite clear that there remain major gaps in the knowledge of ecology and behaviour ot this fly. In an attempt to bridge these gaps, the ICIPE is focussing attention to specific areas which, the institute believes, are crucial for environmentally safety and that meaningful control measures can be succesfully realized.

These include:
(a) Development of trapping techniques capable of being utilized in control strategies
(b) Development of population model for the main vector of Trypanosomiasis <u>Glossina pallidipes</u> which can be extended to an epidemiological model, and
(c) Utilization of natural enemies of tsetse, i.e. virus to the control of tsetse flies

There has been very good progress in these areas. A high resolution population model is operational and is being used to check methods of analysis and identity where more data are required. Many of the parameters required for the epidemiological model have been estimated. New traps which are relatively cheap and can be constructed by the local community have been tested successfully.

The ICIPE's research on developing an integrated tsetse control approach is based on the use of tsetse traps, new potent attractants, and tsetse pathology, whitin known tsetse population patterns. The ICIPE has chosen to work on <u>Glossina pallidipes</u>, the main tsetse vector of human and animal Trypanosomiasis in Eastern and Central Africa. The species occurs in isolated or semi-isolated foci, which makes its management possible under defined ecological conditions. The ICIPE's tsetse population monitoring methology before, during and after spraying of Lambwe Valley, in Western

Kenya, has shown the ineffectiveness of the control solely through spraying insecticides. Its research in Nkruman Escarpment in Masailand has compiled a data bank on tsetse behaviour, biting cycles, and insect parasite relations, which can be used to build a tsetse population model. Furthermore, it has improved efficiency of traps discovered potent, long-range attractants which can be incorporated in the traps for use by rural communities to control tsetse populations, It is proposed to test this control strategy in a pilot trial scheme in Nkruman in Rift Valley Province of Kenya.

At Nkruman, more or less equal emphasis is devoted to the vector on the one hand, the diseases on the other , and each member of the scientific team is aware of the activities of the others. More importantly still the discussions between experts mean that the research programme of each can respond to the findings of the others. We have waited 80 years for this to happen, and current progress at Nkruman is so rapid and impressive the project should therefore continue on more or less the same lines, with some modifications, before an all-out attempt is made to control the tsetse population. An emphasis on eradication would make the ICIPE project like many others throughout the continent.

The Modelling Approach

Whenever control or eradication of tsetse flies is attempted, either an additional adult mortality is added to existing natural ones (e.g. by insecticide application) or the reproductive rate is decreased (e.g. by the sterile male technique). If however density dependent mortality is acting at any stage of the life cycle, the reduced densities after control will result in a relaxation of such mortality. Thus the population may stabilise at a lower equilibrium if control is maintained, or increase again rapidly when control measures cease. In order to develop cost effective control or eradication strategies, we need to understand such natural regulatory forces. As Rogers (1975) pointed out however, we cannot claim to understand these unless we are able to construct a realistic mathematical model that is able to simulate the behaviour of field populations.

One of the main objectives of the ICIPE Nkruman Tsetse and Trypanosomiasis Project is to develop such a predictive population model. This has necessitated three years of intensive field work on both pupal and adult stages of G. pallidipes at Nkruman, Kenya. Results so far suggest that the most important density independent mortality factors are saturation deficit and flooding of the larviposition sites. Density dependent mortality appears to be acting at both the pupal stage (predation) and the OB adult stage (starvation or emigration resulting from competition at the host). A high resolution model with age structure running on a physiological time scale is now operational, and is at the stage where parameters estimated from field data are being incorporated. Although the model is still in a relatively simple form, it is already giving information on the parameters to which the population is most sensitive as well as focussing attention upon areas where data are lacking. As its predictive power improves, it will be possible to test various control strategies on the model.

As well as modelling the vector population dynamics, the project was designed to provide sufficient data for a full epidemiological model, of which the population model would be a component. Thus a multidisciplinary team approach was adopted with trypanosome infection rates in both the flies and cattle being monitored. Many of the required parameters have been estimated, with research at present orientated towards cattle grazing patterns and wild host infection rates.

Community Participation in Control

The emphasis at Nkruman has been put on developing community participation in the proposed control programme with a maxximum of external inputs. Over the last twenty years in Africa, reliance has been put on high technology control techniques often funded externally. These have frequently proved to be unsuccesful or only short term palliatives because of logistical difficulties and the problem of reinvasion. By involving the local community fully and utilising the new low cost trapping technology we believe these problems can be overcome. Initial field testing of new traps and odour baits at Nkruman are forming part of a population manipulation programme which enables us through the model to test the resilience of the population . More extensive deployment of the traps should start soon.

So far the most cost-effective bait tested has been acetone (500 mg/h) and cow urine which have given average increases of 13 x for males and 17 x for females when baited and unbaited biconical traps are compared. The index of increase is however very dependent on temperature. A simplified version of the Zimbabwe F3 trap costing about Kshs.100.00 which can be constructed locally in the Masaii manyattas has further increased female catch by 1.5 - 5 x over the catch of a biconical trap.

By combining a relatively sophisticated research approach (population modelling) to understand population processes, with a low cost community participation control strategy (odour baited traps) we believe that cost-effective tsetse control can be achieved.

The IEMVT's present and future programmes for Glossine control

G.Tacher
IEMVT, Maison Alfort, France

This summary of the IEMVT's (Institut d'Elevage et de Médecine Vétérinaire des Pays Tropicaux) glossina control activities does not reflect all of French activity in this field. Other bodies are also working directly or indirectly on this problem, the main ones being the Ministry of Cooperation, ORSTOM (Office de la Recherche Scientifique et Technique d'Outre Mer) and the universities.

The IEMVT, which has been working on Central and West African glossina since the late Fifties, began its research with ecological studies of the various glossina species in the Central African Republic (CAR) and Chad (Central Africa) and in Senegal (West Africa). It was on the basis of this work that detailed distribution maps were drawn up for these three regions and anti-tsetse campaigns involving ground spraying of insecticides devised and implemented.

There were two such operations in the CAR; the first, on a small scale (8 000 ha), was in 1961 and covered two <u>fusca</u> group species, whereas the second, a larger operation (45 000 ha), was carried out in 1962 against <u>G. fuscipes fuscipes</u>. As for Chad, a vast spraying operation against <u>G. tachinoides</u> took place from 1972 to 1974 in the Lake Chad basin, this time covering more than 250 km of the banks of the Chari river and its tributaries. This operation was a complete success thanks to precise knowledge – contained in Jean Gruvel's doctoral thesis – about the distribution and biotopes of this species.

The first attempts to breed glossina in Njamena (Chad) and Bouar (Bewiti) in the CAR date from that time, and it was about then that the idea of a biological control study came about. The first project was proposed in 1968 or thereabouts and, for the record, it was the EEC which financed the purchase of the first irradiator. This irradiator was sent to Bangui in 1970 but events in the CAR at that time meant the project had to be abandoned.

117

However, studies continued at our HQ in Maisons-Alfort with a view to
rationalizing glossina breeding techniques, mastering all aspects of
production and determining the optimum doses of gamma rays needed to combine a
high sterilization rate with maximum longevity and resistance. From 1974
onwards these findings were applied at the Centre de Recherches sur les
Trypanosomiases Animales (CRTA, Animal Trypanosomiases Research Centre) at
Bobo-Dioulasso, which had been set up to develop the control method of
releasing sterile males and adapt it to field requirements. This project soon
became a trilateral affair linking Burkina, the GTZ (Gesellschaft für
Technische Zusammenarbeit) and the IEMVT.

Our knowledge of the biology of glossina in West Africa and about the breeding
of flies in tropical surroundings has benefited from several research
projects. Marking and recapture operations involving two species of riparian
glossina (G. palpalis gambiensis and G. tachinoides) made it possible to
establish the biotope, longevity and dispersion capacity etc. for both wild
insects and sterile males. This enabled us to determine when and where the
sterile males could best be released and to provide barriers accordingly.

The glossina breeding stations were considerably rationalized and simplified
so that by using specially designed equipment it was possible to reduce
personnel while increasing production. Another major improvement was the
initially partial then total changeover to artificial feeding on
slaughterhouse blood by means of a membrane, which meant a great reduction in
the sensitive issue of breeding host animals (rabbits and guinea-pigs). All
these improvements led, for the first time in Africa, to mass-breeding of
glossina, reaching the 350 000 reproductive female mark. As a result of these
measures the cost of producing sterile males became competitive with that of
traditional control methods.

Application of all this research work in a development project in the
Sidéradougou region made it possible to get to grips with the tsetse fly, and
to eradicate it from an area of 3 500 km^2, by combining repeated releases of
sterile males with the use of traps and insecticide screens - a technique
first developed by the CRTA - in order to reduce the indigenous wild
population and prevent reinfestation.

At the same time a number of other projects were able to take advantage of the
fly breeding facilities, including projects by the WHO (glossina dispersion),

ORSTOM (sensitivity to insecticide) and the LAFIA project in Nigeria (despatch of G. tachinoides pupae).

Although eradication has been achieved, the question now is to prevent reinfestation; this is not only a biological problem but also has human, administrative and even political implications, which I believe are just as important as the biological ones.

The present project is coming to a close, and Burkina, the GTZ and the IEMVT are now wondering what is to become of the breeding facilities (over 300 000 females from 3 species, existing capital investment, trained staff, know-how).

The other main approach to tsetse control does not involve sterile males but only traps and screens, with or without attractants, possibly in the form of pheromones.

At present the CEC (Commission of the European Communities) is providing 50% of the funds for a project being carried out together with the GTZ and — in a new phase — with Dr Vale of Zimbabwe, the TDRI of London and the TRL of Bristol. This project is also being coordinated with a GTZ project in the Ivory Coast.

The French contribution to this project involves using the breeding facilities in Maisons-Alfort to study pheromones in collaboration with the CNRS (French National Scientific Research Centre). All the field experiments in Africa are carried out at Bobo-Dioulasso, and the French Ministry of Cooperation has provided funds to support this project in 1986.

Thus, the Bobo-Dioulasso centre is West Africa's counterpart to the East African centre in Zimbabwe.

This helps strengthen links with East Africa, as does the IEMVT's involvement in the EEC-funded East Africa project, which Mr Lovemore has spoken about. Our role in the project is to monitor the environment, with Saarbrüecken University acting as team leader.

Other projects more specifically geared to development are being devised, though the money to fund them has not yet been found.

Our own publication and its special issues help spread information in this field. The IEMVT also has a hand in devising the TTIQ.

Another facet of tsetse control activity is training - which is becoming increasingly necessary - using methods involving the local population.

As far as training middle-level staff is concerned, the ELAT is just about ticking over, being used for nothing more than refresher courses.

As for training senior staff, over the past 10 years or so the IEMVT has arranged a trypanology course once every two years. The next one was scheduled for 1986, but has been cancelled owing to lack of funds. These jointly funded courses lasted six months and involved three months theory and three months practical fieldwork. The English-language course is still continuing, I believe.

The problem is knowing whether the French-language course should continue as it is, or whether it should be modified or even condensed to correspond more realistically to the funds available.

I would like to end by noting that the IEMVT has always adopted a multi-disciplinary approach to trypanosome control, because it believes that progress will be made through an amalgam of methods. Under the Sideradougou control programme this involved the use of several complementary methods (traps, screens, sterile males). Other aspects of trypanosomiasis studied include trypanocides (relatively few in number) and trypanosomiasis immunology, but the main stress is on trypanotolerance. The study is largely carried out in conjunction with the GTZ and covers immunology and genetics, with the emphasis on tracers.

The IEMVT's main thrust, therefore, involves traps and screens linked to attractants as well as immunological and genetic studies of trypanotolerance.

Like the speakers before me, I hope that the control of trypanosomiasis - which is still the main pathological obstacle to livestock breeding in Africa - will benefit from coordinated efforts, and hence increased cooperation between all bodies concerned.

BIBLIOGRAPHY

- Annual reports from the Central Africa research region - Farcha laboratory in Chad and Bouar laboratory in the Central African Republic; since 1955.

- Revue d'Elevage et de Médicine Vétérinaire des Pays Tropicaux, four issues per year since 1955.

- Les moyens de lutte contre les trypanosomoses et leurs vecteurs - Control programs for trypanosomes and their vectors. Acte du Colloque (Colloquium proceedings) PARIS 12 - 15 March 1974, 387 pages. Institut d'Elevage et de Médicine Vétérinaire des Pays Tropicaux - Ministére de la Coopération. International Office of Epizootics.

- Glossines et Trypanosomoses: Revue d'Elevage et de Médicine Vétérinaire des Pays Tropicaux. Special edition 1984, 315 pages.

ODA's present programmes and future aims for tsetse control

T.Jones, R.Allsopp & T.J.Perfect
Tropical Development and Research Institute, Overseas Development Administration

Summary

ODA supports multivarious research and development programmes in the
field of tsetse and trypanosomiasis through its own R & D programmes,
bilateral aid projects through multilateral aid agencies and in collaborative ventures with international organisations. These cover
laboratory investigations, field trials and control operations
embracing chemotherapy, trypanotolerance and especially vector
control by chemical and non-chemical techniques, some of which it has
pioneered, developed and refined over many years, in a keen awareness
of the need to maintain the quality of the environment. The aim is
to exploit all major forms of tsetse control techniques in harmony
and so devise control strategies which meet all the necessary
technical, logistical, economical and environmental requirements of
specific target areas where tsetse control in the context of sound
land-use planning can lead to orderly and rational agricultural
development.

1. Introduction

Since the dawn of the 20th Century and our earliest realisation of
the role played by tsetse flies in the transmission of trypanosomiases, a
succession of British Governments, many national institutions and
countless interested individuals of UK origin have become involved in the
seemingly endless search for a means to eradicate these dreadful and
ubiqitous diseases. Recognising the social and agricultural development
of many African countries, recent UK Governments, including the present
one, have, through the offices of the Overseas Development Administration
(ODA), and its predecessors such as the Commonwealth Development Fund
provided manpower and finance for activities which ramify the entire
international fabric of tsetse and trypanosomiasis control.

Because of the pragmatic way in which the tsetse problem has been
approached over the years, work on various aspects of it has been pursued
by a number of separate institutions and through various forms of ODA
support and finance. The result is a somewhat scattered distribution of
resources and efforts which may be contrary to the preferred approach in
other fields but which, as experience has shown, greatly facilitates a
truly multidisciplinary effort and continuity of purpose. ODA is now able
to draw upon the strengths of several institutions which are highly
regarded in their field and where there has been need, ODA has created
capabilities or increased them at institutions to enable them to address
more specifically the problems of tsetse and trypanosomiasis. ODA has for
example its own Scientific Units in the form of TDRI (Tropical Development

and Research Institute) and LRDC (Land Resources Development Centre), and 'Associated Bodies' such as the Tsetse Research Laboratory (TRL) at Bristol and the Centre for Tropical Veterinary Medicine (CTVM) at Edinburgh. The core budgets of these UK based institutions and finance for their programmes of research and development are provisioned and administered by ODA's Natural Resources Division as an element of bilateral aid. The output from these institutions and many others which receive ODA grants for specific research, development and training activities is generally translated into practical assistance to African countries through country specific bilateral technical cooperation programmes administered by ODA Geographical Departments and through ODA funded R & D Projects designed to further advance knowledge and technology for the benefit of regions as a whole. Both bilateral TC and Regional R & D programmes in tsetse control provide above all else perhaps teams of experienced specialists whose ultimate objective is to create local capability and capacity for the conduct of tsetse research and control activities.

Britain's contributions to multilateral aid, including that allocated to tsetse and trypanosomiasis problems is channelled through such organisations as the World Bank Group, FAO, WHO and the EEC. For instance ODA as a partner in the European Community's aid programme to African, Caribbean and Pacific (ACP) countries, supports such programmes as the Regional Tsetse and Trypanosomiasis Control Programme for Southern Africa (RTTCP). The Consultative Group on International Agricultural Research (CGIAR) also receives ODA support. CGIAR funds the International Laboratory for Research in Animal Diseases (ILRAD) and the International Livestock Centre for Africa (ILCA) both of which are committed to the elimination of trypanosomiasis.

The various channels through which aid and support are given by ODA are summarised in Fig.1 and the programes supported by specific allocations are summarised below.

2. Summary of ODA Support for Tsetse Work
2.1 The Scientific Units
2.1.1 Tropical Development and Research Institute (TDRI)

The College House wing of TDRI (formerly COPR) has been principally concerned over many years with the development of techniques for field scale tsetse control. Its initial involvement with WHO in helicopter spray trials in Upper Volta was followed in 1973 by studies in Botswana on ultra-low-volume techniques (ULV) for insecticide applications which also embraced detailed studies on the ecology of G.morsitans centralis and a 16 man year study on the environmental impact of tsetse control operations.

Since 1981 TDRI with ODA funding has conducted a major exercise in Somalia to examine the feasibility of eradicating the isolated tsetse populations in the Shebelle and Juba riverine areas using fixed wing aircraft to apply sequential aerosol aerial spraying (SAAS). Parallel studies have been conducted in Zimbabwe on airspray application.

Meanwhile since 1980 the Insect Products section of the Grays Inn Road wing of TDRI (formerly TPI) had at the request of the Government of Zimbabwe begun identifying the components of those host odours which were known to be attractive to tsetse flies. A collaborative project with TRL, (which supplied flies and bio-assay facilities) led to the identification of further components which also proved attractive to the Zimbabwe tsetse species. Similar research at TDRI on West African tsetse species has been supported by WHO through a collaborative project with TRL, Zimbabwe, Congo, Ivory Coast and Burkina Faso.

In 1979 COPR, in financial partnership with FAO and WHO and with inputs in kind from the OAU and the IEMVT initiated the publication - Tsetse and Trypanosomiasis Information Quarterly (TTIQ).

2.1.2 Land Resources Development Centre (LRDC)

The general increasing awareness of the need for sound development of land areas freed from tsetse infestation has led to LRDC becoming even more deeply involved in ODA's tsetse control programmes and it contributes the vital element of land use planning in ODA's integrated approach to the tsetse problem. In Somalia for example LRDC has a major input into the National Tsetse and Trypanosomiasis Control Programme.

2.2 Associated Bodies
2.2.1 Tsetse Research Laboratory (TRL)

The TRL as an independent unit of the Bristol University School of Veterinary Medicine at Langford was founded in 1961 to create and maintain productive colonies of tsetse flies in captivity, and conduct research into tsetse biology, control and disease transmission. Its success in developing the membrane feeding technique and hence a capacity for rearing tsetse opened up opportunities for research hitherto denied to temperate laboratories.

Apart from its own research on tsetse and trypanosome biology and physiology, it also has collaborative programmes with other institutes and overseas governments on attractants, sterilization and trapping techniques and its programme which allows more speculative approaches to vector and disease control effectively complements that of TDRI and other laboratories.

2.2.2 Centre for Tropical Veterinary Medicine (CTVM)

CTVM under ODA research grants conducts research projects on Trypanosoma congolense and T. evansi. The first is concerned with immunology and tissue culture studies in the pursuit of drug regimes and vaccine production. The second project will improve knowledge of epidemiology to allow improved methods of control in cattle, buffaloes and camels.

2.3 Other institutions receiving research support
2.3.1 Imperial College at Silwood Park

The Department of Pure and Applied Biology is conducting a three year ODA commissioned research project to study the mating behaviour of Glossina with particular reference to pheromone-baited sterilising traps (ie. in conjunction with TRL and a TCO in Zimbabwe) using video recordings for field and laboratory studies of tsetse behaviour. This work pertains to the new development of trapping and chemosterilisation control methods which appears to offer some promise for tsetse control.

2.3.2 Cambridge University

The ODA financed project to study over six years the biology and genetics of pathogenic trypanosomiasis has recently ceased and results are awaited. Meanwhile a smaller research project has started to investigate further the genetic systems of one trypanosome species. This highly complex study involves the screening of mixed procyclic (bloodstream) cultures for the occurrence of hybridisation between trypanosomes, initial development of plasmid vectors designed to mediate DNA-dependent genetic transformation of trypanosomes, and the development of new genetic markers for trypanosomes based on restriction fragment length polymorphisms. This work mostly pertains to the expression of variable surface antigens and hence interaction with the mechanisms of immunity.

2.3.3 Glasgow University

The Departments of Physiology and Medicine at Glasgow University have close working links with ILRAD, ILCA, TRL, CTVM and ITC (International Trypanotolerance Centre). Their ODA commissioned research involves investigations of the factors influencing trypanosome virulence in susceptable and trypanotolerant animals and studies in cattle to examine factors influencing the duration of chemoprophylaxis. The University is also working on the identification of surface antigens of bloodstream trypanosomes which is fundamental to understanding the immune mechanisms of trypanosomiasis.

2.4 Geographical Desk Projects

Most of the activities described above in terms of UK institutional efforts are largely or wholly financed from ODA research support funds (R & D contracts and Scientific Unit/Associated Body core support), but ODA also finances research and development work through its country Technical Cooperation (TC) programmes, often as part of integrated TC/ R & D/core funded programmes.
Current Geographical Desk projects are as follows:-

2.4.1 Gambia: International Trypanotolerance Centre

Three technical cooperation officers (TCOs) are being provided, supported by equipment and consultancies financed from Special Initiative funds over three years. The project aims to assess the productivity of trypanotolerant cattle in tsetse-infected areas and to develop low-cost methods of increasing productivity. Close links are maintained with Glasgow University.

2.4.2 Zimbabwe: Olfactory Attractants and Traps

Two TCOs are employed in Zimbabwe to develop more effective insecticide/target combinations and one at TRL to investigate the effects of host odour on flight behaviour. Large scale field trials of olfactory attractants and insecticide impregnated targets are being carried out. There are related R & D projects on mating behaviour at Imperial College Silwood Park - and development of automated tsetse fly traps. TDRI core-funded research is also relevant.

2.4.3 Zimbabwe: Monitoring of Aerial Application of Insecticides

TC financed visits have been carried out by TDRI staff to monitor the efficiency of the aerial spraying programme. Related activities are a

TDRI/R & D financed research programme to improve efficiency in rugged terrain and the EEC project to eradicate tsetse in Southern Africa (RTTCP).

2.4.4 Zambia: Manpower Assistance

Two OSAS entomologists are presently with the Department of Tsetse Services.

2.4.5 Somalia: National Tsetse and Trypanosomiasis Control Project

As mentioned above, both TDRI and LRDC have been extensively involved with this project. Six TCO's were employed on tsetse surveys and control and a number of both short and long term TC officers were assigned through LRDC, for land use planning.

3. ODA's Recent Achievements and Present Priorities in Tsetse Control

ODA funded research has made significant scientific contributions to the field of tsetse and trypanosomiasis control in entomology, proto-zoology, immunology, therapeutics and vector control, but it is to the last mentioned aspect that ODA has most recently committed its greatest support.

Vector control, predominantly with the use of insecticides has impressive historical credentials. It provides immediate, visible and affordable results, and is therefore, understandably, the method most widely practiced by African Governments. It is likely to remain so regardless of some multilateral or bilateral insistence to the contrary and therefore fully justifies any effort to improve its efficiency and reduce its unwanted environmental effects. It must as is now generally acknowledged be undertaken in close collaboration with medical, veterinary and land-use specialists and with a full awareness of potentially far reaching socio-economic or political implications. The effects on non-target organisms, including man, his domestic livestock and other protein resources such as fish are of particular significance. The Scientific Units of ODA and Associated Bodies have unique experience of these many facets and are able to provide competent multidisciplinary support to vector control operations and research. They have helped to refine the methods used for the insecticidal control of tsetse flies and these are now far more environmentally acceptable than they were not so very long ago. Aerial applications of low dosage aerosols have greatly reduced the amounts of insecticide applied to a target area and the development of impregnated traps or targets has made the placement of insecticides even more specific and environmentally benign.

Despite great improvements in the environmental acceptability of insecticidal methods, ODA continues in response to developing country requests to investigate the non-target effects of both recent methodologies and the more traditional methods such as ground spraying with DDT, which are still highly cost effective, still in use and are still the preferred or only financially feasible answer to the problems of some tsetse afflicted countries.

The major thrust of ODA's present and recent contributions to tsetse and trypanosomiasis control manifests itself in three distinct forms viz.

(a) tsetse control operations

(b) research and development to advance knowledge and technology with particular reference to the use of aerial spraying and odour baited, chemically impregnated traps and targets

(c)   environmental monitoring studies associated with both control operations and research and development programmes.

3.1 Present and Recent Involvements with Operational Tsetse Control

In Somalia where tsetse flies are confined to the relatively narrow riverine areas of the Shebelle and the Juba and are isolated from all other populations the prospects of eradication have long seemed attractive. In 1980 the Government of Somalia with the assistance of the Arab Fund for Economic and Social Development (AFESD) and ODA established the National Tsetse and Trypanosomiasis Control Programme (NTTCP). In the intervening years this programme staffed by local officers and TDRI and LRDC staff has surveyed and mapped the fly populations and undertaken control trials in the Middle and Upper Shebelle valley to test the feasibility of an eradication programme. Whilst meteorological conditions are extremely demanding airspray trials have shown that Sequential Aerial Aerosol Spraying (SAAS) can be very effective and results justify further extensive trials over two years involving both SAAS and various ground based control systems including targets and traps designed to refine techniques, identify the parameters for using these techniques and develop an effective strategy for any subsequent eradication campaign.

The Zimbabwe Government has used aerial spraying techniques to progressively roll back the tsetse populations along the southern shores of Lake Kariba and in doing so has learned some of the problems and limitations of these methods - particularly with relation to the highly variable topography of the area. TDRI has been closely associated with this four year programme which has provided important guidelines for the EEC Regional Tsetse and Trypanosomiasis Control Programme for Southern Africa (RTTCP) and will continue its own development programme with EDF support within the RTTCP.

TDRI, or more correctly its predecessor COPR, was involved for many years with the development of aerial spraying and environmental monitoring in the unique Okavango wetland of Botswana. Two major TC projects were carried out between 1976 and 1980 to study the ecology of Glossina morsitans centralis as it relates to control and to comprehensively assess the non-target effects of aerially applied endosulfan. This latter study was largely responsible for the present acceptability of the Sequential Aerosol Aerial Spraying (SAAS) technique and has established standards by which future monitoring programmes will be assessed. TDRI staff are currently assisting the Canadian Government to define future tsetse control requirements in Botswana and the neighbouring countries of Angola and Zambia.

3.2 Research and Development
3.2.1 Aerial Spraying

The sequential application of low dosage aerosols from fixed wing aircraft (SAAS) is a highly technological, self contained method of controlling tsetse. It is not labour intensive and has the potential to eliminate tsetse from vast areas in a short space of time.

Its operational parameters are still, however, inexactly defined. Current research on aerosol characteristics concentrates on the relationship between droplet size, numbers and both type and dose of active ingredient. (A suspected requirement for small droplets of high toxicity may limit the use of the currently favoured insecticide and research on alternatives is essential).

The behaviour of droplets in relation to formulation, generation, environmental variables and impaction on flies is being studied using wind tunnel techniques. This laboratory work interacts with field testing and the monitoring and evaluation of practical control operations designed to establish how the prescribed parameters can be met under operational conditions to increase the robustness of the technique.

The major objectives of the field programmes is to establish the limits for effective use of SAAS in relation to terrain, devising alternative methodologies where appropriate and examining the interface of SAAS with other control techniques such as the use of traps and targets. The latter area is a major research need currently constrained by manpower resources.

The effectiveness of SAAS depends to a large measure on timing. Spray cycles must be repeated at an interval that precludes mature females from larvipositing. The rate of development of larvae is sensitive to environmental conditions, particularly temperature, and further research is to be initiated on variation of the larval development rate within and between species to improve the timing of spray cycles.

SAAS will exploit the rapidly increasing development of sophisticated electronic equipment. This may, for instance, include the most recent computerised terrain profile matching system of track guidance, presently developed for fighter aircraft and missiles but already under investigation for tsetse control. It may also include satellite aided global positioning systems (GPS) when satellite coverage of Africa is available - probably within the next two to three years. Forward looking infra red night vision equipment (FLIR) is also being considered to improve the safety and efficiency of these, mainly nocturnal spraying operations. In these respects, SAAS differs from, but compliments the "appropriate technology" of baited trapping methodologists.

3.2.2 Odour-baited, Insecticide Impregnated Targets and Traps

There has recently been a resurgence of interest in the use of stationary traps and similar structures which were highly successful and popular for tsetse surveys and, to a lesser extent, control, in the 1930's and 40's - ie. before the dramatic introduction of insecticides such as DDT. Pioneered by the French scientists Challier and Laveissiere in West Africa and Vale in Zimbabwe, a variety of traps have been developed and have proved most successful as survey tools. For control their visual attraction alone was found to be insufficient but under the guidance and innovative flair of Professor Bursell (now at TRL) and Dr Vale in Zimbabwe, the search was instigated for ox-like odours which could be used to enhance the attraction of these stationary structures. Expertise from TDRI and the TRL was enlisted to hasten the discovery of those components of ox and porcine odour which were most attractive to tsetse flies.

Research on odour attractants, through analysis of host odours using Gas Liquid Chromatography, electroantennography and mass spectrometry at TDRI, has resulted in the identification of chemicals such as 1-octen-3-ol and 2-butanone which have greatly expanded the potential for trapping and killing tsetse flies. Eradication has been achieved in an island habitat and seems within reach on a larger scale. Results in a 600 sq km trial in Zimbabwe are very encouraging.

129

The practicability of large scale use of traps and targets is profoundly affected by the density at which they must be deployed. This in turn depends on their attractive range and thus improvements in the chemical bait are of key significance. There is good evidence that other components remain to be identified which will further enhance the attractiveness of the currently used odour bait. The principal component characterised for tsetse species common in southern Africa appears to be ineffective for the West African G. palpalis and the continuing search for further attractants must be given high priority.

Operational and logistic problems have been encountered when deploying odour baited targets on a large scale but their development continues within the EEC RTTCP and having established their basic capability to eradicate tsetse flies from large areas they can be incorporated into integrated control strategies. Indeed, such is their potential, particularly with regard to economics and the environment, that their development should be seen as vital to the realisation of cost effective and internationally acceptable integrated tsetse control methodology.

3.3 Environmental Monitoring

As a signatory to the resolutions of the Stockholm Conference on Man and the Environment, Britain is committed to respecting environmental quality and has thoroughly pursued its obligations by comprehensively monitoring any of its own tsetse activities which might have environmental consequences. An Impact Assessment Group has been established at TDRI and will be continuing the DDT studies first carried out in Zimbabwe in 1984. TDRI staff will also be seconded to the Environmental Monitoring Group commissioned by the EDF and supervised by the Institute of Biogeography, Saarbrucken, which will monitor all control methodologies employed within the EEC RTTCP.

4. Conclusions

The benefits to be gained from successful and permanent eradication of tsetse and trypanosomiasis are well documented and inherent in the EEC's involvement particularly on so large a scale as that of the RTTCP. The realisation of these benefits will be dependent upon investment in and the maintenance of economic and social infrastructures in cleared areas plus the availability of populations willing to settle and seasonally graze their cattle in such areas. To this end tsetse control must be implemented in conjunction with sound land use planning although it should also be realised that land use may also include wildlife and wilderness areas where pockets of infestation could not remain without risk. Subject to these conditions, tsetse control can assist sub-Saharan Africa to sustain and feed its population through improving the quality of its natural resource base, and more cost effective control can transform previously marginal agricultural programmes into viable investments. The attainment of cost effectiveness must always be inherent in tesetse control programmes if only to ensure continuity beyond the donor funded life of any programme: it is certainly a major consideration in ODA funded projects. The current cost of aerial spraying varies between locations but is in the region of US$350 per square km. This is approximately 60:40 flying charge and insecticide costs. The common fly belt to be cleared in the RTTCP is 320,000 square kilometres and it is fairly evident that even small improvements in operational technique to reduce flying time or reductions in dosage rates could result in immense savings. Similarly more effective odours would result in lower trapping densities and correspondingly lower costs.

The immediate future for tsetse control is dominated by the exciting prospect of significant progress within the RTTCP. The provision of funds and facilities on such a scale will enable institutions such as TDRI to investigate 'the most modern' electronic aids that, even for trials, would be beyond the scope of any single, unaided institution. It will also hasten the discovery of new attractive odours and the operational refinement of the impregnated target concept.

The scale of this programme should also ensure that target areas are sufficiently large and progress adequately sustained to reduce, or perhaps eliminate, the ever present constraint to eradication which is reinvasion. The scale of this operation also shifts the emphasis from control in the SADCC countries to eradication and should thus ultimately eliminate the recurrent costs which have been estimated at £6m pa. Taking account only of the benefits which would accrue from increased livestock production within the SADCC from a ten year eradication programme at a cost of £100m, the estimated internal rate of return (IRR) is 13.2%.

Looking beyond such immediate programmes as the RTTCP we should recognise the intrinsic problems of tsetse control and attempt to alleviate some of these while the opportunity is present. For instance, there is a serious shortage of skilled manpower at the more senior levels of tsetse and trypanosomiasis control departments. Organisational ability for large logistically complex control programmes is lacking as is expertise in the field of applied research. Also tsetse fly infest remote, inhospitable areas thus presenting immense logistical problems and discouraging staff from working in the spartan isolated conditions.

Efforts can be made within such programmes as the RTTCP to alleviate these problems. For instance, counterpart training should be given a high priority. Control strategies must utilise the most effective methodology in specific areas - which is the fundamental principle of integrated control - but should also take account of operator requirements which may mean persisting with a less suitable technique or perhaps providing access and facilities.

Human and cattle populations continue to expand in Africa and will inevitably exacerbate the problems of trypanosomiasis unless drastic and extensive steps are immediately taken. European governments and institutions have shown particular awareness of this problem and have mobilised their extensive collective experience to exploit high and low technology methods of vector control whilst retaining a sound operational base in such fields as trypanotolerance and chemo-therapeutics. Ultimate success in the effort to eliminate these diseases will unquestionably depend upon the integration of several techniques and this in turn will require the cooperation of many - historically and characteristically autonomous - scientific institutions. This latter requirement has already been met to a considerable degree in recent times and collaboration is noteworthy. The channelling of multidisciplinary expertise through a federation of European countries to aid a federation of African countries is similarly laudable and perhaps within this spirit of friendly cooperation, tempered with scientific achievement, lies the seeds of defeat for a scourge which is almost anachronistic in the context of present global technological capabilities and which must finally and irrevocably be expurgated.

Fig.1.

ODA Inputs to Tsetse and Trypanosomiasis Research and Control

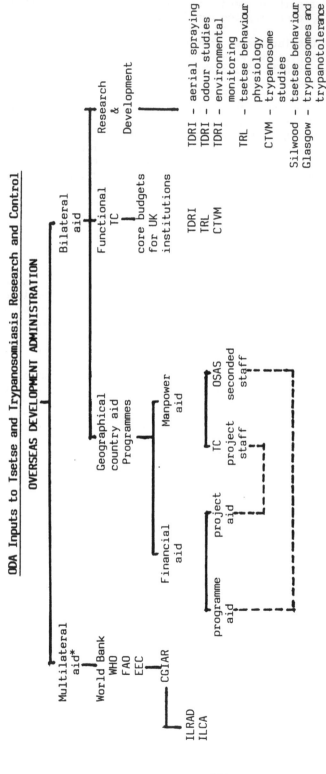

[* largely non-specific but some funds could relate to tsetse and trypanosomiasis research or control]

132

The FAO programme for the control of African animal trypanosomiasis and related development

J.G.Le Roux

Animal Production and Health Division, Food and Agriculture Organization of the United Nations

SUMMARY

The FAO proposal for a long-term Programme for the Control of African Animal Trypanosomiasis and Related Development was presented to the World Food Conference in November 1974. A recommendation was adopted that the Programme should be implemented as a matter of urgency and should receive high priority in the FAO programme of work and budget. Following recommendations of support by FAO statutory bodies the preparatory phase was launched which led to implementation of a large-scale programme in 1980.

I. Preparatory phase (1975-1979)

The objectives of the preparatory phase were to establish the necessary technical and administrative structures at international, regional and national levels; intensify the training of personnel at all levels; conduct applied research, surveys and field trials in order to improve control techniques; promote research into trypanotolerance and establish multiplication centres; assess the socio-economic implications of trypanosomiasis and its control including land use after tsetse elimination; mobilize the funds and services required to fulfill the objectives which have been set and to formulate plans for control and development programmes.

The need to include land use planning and area development in tsetse and trypanosomiasis control programmes was recognized at an early stage as was the need for monitoring the impact on the environment.

To achieve the above objectives a number of activities were implemented, namely:

- Establishment of a Trypanosomiasis Control Unit at FAO Headquarters as well as administrative structures for the coordination and management of the Programme.

- Seven expert consultations were organized to define a detailed work plan for the preparatory phase, to review the general problem of human and animal trypanosomiasis, to determine modalities for implementation of the Programme and to secure the technical data base concerning trypanotolerance, African animal trypanosomiasis and the economics of its control.

- Three joint FAO/Industry Task Force meetings were convened to ensure reciprocal technical feedback.

- Consultancy missions were fielded in 27 of the 37 affected African countries to advise on the establishment of national control services and to assist in project formulation. In addition, the economic impact of trypanosomiasis on rural development was studied in 16 countries.

- Control projects were implemented in Mozambique and Niger and a tsetse distribution survey was undertaken in Ivory Coast in collaboration with GTZ.

- A jointly sponsored tsetse and trypanosomiasis information service was commenced; reviews of the current state of knowledge of insecticides and application equipment, and the impact on the environment of tsetse control operations were commissioned and published. A study of the status of trypanotolerant livestock breeds was carried out in collaboration with the International Livestock Centre for Africa and the United Nations Environmental Programme.

- Following a consultancy mission for the establishment of a regional training centre in Eastern Africa, a project was initiated for training middle-level personnel in Zambia. A total of five joint training seminars were conducted in which 142 personnel of the leadership cadre participated from both anglophone and francophone countries.

- A total of US$100 000 was allocated to seven African research institutes and three laboratories based in Europe to promote and assist research into tsetse ecology, control by insecticides and the effect on the environment, arboricides for bush control, insect growth regulators, epizootiology of trypanosomiasis in bovines, ovines and camels, trypanosomiasis diagnostic techniques, trypanocide pharmacokinetics and trypanotolerance.

- Mobilization of resources and services by personal and consultancy visits to assistance agencies to secure cooperation and collaboration in Programme implementation.

The Commission on African Animal Trypanosomiasis was established by the XXth Session of the FAO Conference in November 1979 and related technical and management services were strengthened at Headquarters and in the Regional Office for Africa. In addition an Interdepartmental Working Group, to ensure coordination of activities within FAO, and two Panels of Experts respectively on the Ecological/Technical Aspects and on Development Aspects of the Programme were established in 1980.

2. Programme Implementation

The Twentieth Session of the FAO Conference approved the plans for the implementation of the Programme which started in 1980 with the launching of multi-disciplinary Preparatory Assistance Missions to assist governments in the formulation of development projects taking into account the trypanosomiasis problem and the feasibility of control. FAO also provided assistance to countries to assist in specific technical subjects, some with TCP support. In total 30 countries received assistance for project formulation or implementation and only two of the 37 infested countries had not availed themselves of FAO assistance.

FAO assistance to date has mainly been concentrated on:

Meetings of the Commission on African Animal Trypanosomiasis and Panels of Experts on Development Aspects and Ecological/Technical Aspects of the Programme, as well as consultation on the environmental impact of tsetse control and a workshop on the breeding of trypanotolerant cattle.

Publication in English and French of a Training Manual on Tsetse Control and a Field Guide for Trypanosomiasis Control, seven volumes of Tsetse and Trypanosomiasis Information Quarterly, eight technical documents and a number of consultants' reports on selected aspects of trypanosomiasis control and rural development.

Training of middle- and post-graduate levels of tsetse control personnel bringing the total trained to 62 and 308 respectively from 34 African countries.

Research grants amounting to some US$250 000 on a number of aspects of tsetse and trypanosomiasis control and related development, excluding those aimed at promotion of the Sterile Insect Technique.

Execution of twelve PAG and 28 technical assistance missions, 17 TCP projects, six UNDP-funded projects.

The first Sub-Regional Development Support Unit as well as a Regional Project for the Rearing of Trypanotolerant Livestock were established in 1983 in Burkina Faso, both being financed by the Government of Italy through the FAO/Government Cooperative Programme.

Collaboration has been established with the International Institute for Applied Systems Analysis in Vienna in order to apply the models designed by the population supporting capacity and agro-ecological zones studies for the identification of areas for large-scale and long-term programmes.

In spite of the progress made in implementation of the Programme some delays in resources flow have been observed. In general, the development of the Programme has been seriously affected by changes in the international economic situation. The effects of the rise in oil prices, the subsequent world economic recession of debt servicing have altered the needs and priorities of the African countries. The funding of field activities has also been affected by new problems such as drought and rinderpest, the immediate effects of which were more obvious and many African countries have given priority to emergency actions using funds earmarked for long-term development programmes.

3. Strategy Adjustments

The ultimate goal of effective large-scale tsetse and trypanosomiasis eradication can only be achieved over a long period of time. Maybe future generations will have to work towards this aim as well. Consequently, it is essential to keep this subject under constant review, preparing the ground for large-scale activities through training, research and even pilot schemes.

Initially, the scope of the Programme was pan-African and aimed at involving all the countries affected by trypanosomiasis. This was in line with the wishes of FAO Member Countries expressed in recommendations by FAO

Governing bodies. Considering the funds actually available, it was too ambitious. Consequently, it was agreed that the Programme during the next few years should:

- focus on priority areas, with good prospects for attracting funds for integrated control and development programmes;

- promote simple and low-cost control techniques for areas where the prospects for limited funding are good;

- intensify training activities and allocate funds for research on subjects of immediate importance;

- cooperate with research institutes, concerned organizations and bilateral assistance agencies.

4. Identification of priority areas

The following criteria are usually applied in the selection of priority areas:

- human population density and competition for land

- need to increase food production

- technical and economic feasibility of controlling animal trypanosomiasis in the areas concerned

- government priority for development of the areas

- prospects for funding.

In addition, small or medium-scale trypanosomiasis/tsetse control projects should be implemented in conjunction with selected development programmes.

5. Promotion of control techniques

Trypanosomiasis Control: support will be given to research on selected subjects, strengthening of veterinary services and promotion of chemotherapy and chemoprophylaxis, the last-mentioned through training in simple diagnostic techniques and the strategic use of trypanocides.

Trypanotolerant Livestock: promote the rearing of trypanotolerant livestock, and, where appropriate, wildlife.

Tsetse control by non-conventional techniques: assist the development of insecticide-impregnated traps/screens particularly for control of savannah tsetse species where appropriate.

Training, Information and Research: in training, preference should be given to nationals from countries where facilities and infrastructure exist so that full use can be made of the training received. Preparation of training material, publication of technical material and dissemination of information will continue.

6. Coordination and Cooperation with other Institutions

In order to avoid duplication of efforts and to make better use of available funds and expertise, collaboration will be strengthened with organizations and research institutes such as WHO, UNEP, IAEA, ILRAD, ILCA, ICIPE, OAU/IBAR, Desert Locust Organization, Onchocerciasis Control Programme, and IIASA.

7. Priority areas for integrated large-scale control and development programmes

On the basis of information available, the socio-economic importance and prospects for funding, and considering the preliminary conclusions of the study on application of the agro-ecological zone data to tsetse infested areas, it is proposed that priority attention for integrated large-scale control and development programmes be given to:

The Kagera River Basin (Burundi, Rwanda, Tanzania and Uganda) where there is considerable demographic pressure and need for better utilization of available land. A number of feasibility studies have been completed for integrated development and a preliminary project for tsetse and trypanosomiasis control has been formulated with FAO technical assistance.

The countries covered by the Sub-Regional Development Support Unit in West Africa with particular reference to the Onchocerciasis Control Programme areas.

The common flybelt in parts of Zambia, Zimbabwe, Malawi and Mozambique particularly in respect of training for the manpower requirements.

Consideration will also be given to the common flybelts along the boundaries between Kenya, Uganda and Tanzania where tsetse control activities could be carried out in cooperation with the Locust Control Organization.

8. Preparatory Assistance Missions

Preparatory Assistance Missions will be continued as recommended by the Commission of African Animal Trypanosomiasis. Efforts will be made to carry out more detailed feasibility studies as well as an economic analysis of project proposals and to focus on specific areas scheduled for tsetse control rather than broad national issues.

Joint FAO/IAEA tsetse fly programme: Current and future

D.A.Lindquist
Insect & Pest Control Section, Joint FAO/IAEA Division of Isotope and Radiation Applications of Atomic Energy for Food and Agricultural Development

Summary

During the past decade the Joint FAO/IAEA Tsetse Fly Programme has emphasized the development of mass-rearing and related technologies as a part of the development of the sterile insect technique (SIT) for eradication of various species of tsetse fly. In addition, methods have been developed which can measure sterility induced in field populations resulting from the release of sterile males. Insecticide impregnated attractant devices for suppression and traps for estimating tsetse populations have been adapted for use in Nigeria. Fundamental research has been conducted on the nutrition of the tsetse fly with the objective of developing an artificial diet. A reasonably successful artificial diet is available.

The Joint FAO/IAEA Programme has utilized its in-house research capability at the Agency's Seibersdorf Laboratory, co-ordinated research programmes, and technical assistance field projects to achieve these objectives. The flow of technology into field projects has resulted in rapid technology transfer and successful use of the SIT for tsetse eradication.

During the next several years, primary emphasis will be placed on increasing the efficacy and decreasing the cost of the SIT for use against tsetse flies. Mass-rearing of additional species will be developed and additional field programmes undertaken. Improvements in mass-rearing will reduce the cost of this aspect of the SIT. Additional experience in field programmes will improve strategy for use of sterile males. Tsetse eradication from areas where more than one species exist will be undertaken.

1.   Introduction

The International Atomic Energy Agency was formed as specialized Agency of the a United Nations Organization to foster the peaceful uses of atomic energy. Agriculture was one of the broad subject matter areas selected during the early life of the IAEA. Because there were programmes in the IAEA and the FAO which dealt with the use of atomic energy and agriculture the leaders of these two UN organizations formed the Joint FAO/IAEA Division about 20 years ago. This division is responsible to both FAO and IAEA for atomic energy related agricultural activities. Within the Joint FAO/IAEA Division there are six subject matter sections, one of which deals with the control or eradication of economically important insect pests.

During the early days of the entomology programme of the
Joint FAO/IAEA Division, the practical application of the Sterile Insect
Technique (SIT) was being developed for use against the screwworm fly in
the USA. Since this technology depended upon the use of ionizing
radiation to sterilize the insects, it was natural that this technology
be part of the entomology programme cf the Joint FAO/IAEA Division. In
subsequent years the programme has grown and at the present time emphasis
is placed on the use of the SIT for eradication of the Mediterranean
Fruit Fly and several species of tsetse flies.

One of the more unusual characteristics of the IAEA is that its
charter specifically requires the organization to conduct in-house
research to support its activities in developing countries. This
resulted in the establishment of a research laboratory near Vienna. Part
of this laboratory is dedicated to agriculture, including a significant
entomology research unit. This has resulted in the capability to conduct
R&D in direct support of our field programmes, to investigate some of the
more basic aspects of improving the SIT and to make available facilities
to train individual scientists and technicians from developing countries
in the expertise available at our laboratory.

2.   Species Selection

Since there are many thousands of insect species which cause
economic loss in the production of agricultural commodities, very strict
criteria must be established to select pest insect species on which to
commit resources to develop the SIT as a means of control or eradication.
Our criteria include whether or not previous work has shown reasonable
success in laboratory and if possible field experiments, whether or not
the insect species is of major importance in a number of developing
countries, whether or not area-wide control or eradication is a
reasonable approach for solving the problem and whether there is interest
in developing countries in committing resources to develop new methods of
insect control or eradication. With these criteria in mind, the Joint
FAO/IAEA Division selected the Mediterranean Fruit Fly and the tsetse fly
(several species) as likely candidates.

3.   Current practical use of the SIT

The Sterile Insect Technique is currently used on a practical
basis against the following six species:

(1)   Screwworm (Cochliomyia hominivorax)
(2)   Medfly (Ceratitis capitata)
(3)   Melon Fly (Dacus cucurbitae)
(4)   Mexican Fruit Fly (Anastrepha ludens)
(5)   Onion Fly (Delia antigua)
(6)   Pink bollworm (Pectinophora gossypiella)

In addition, the technology has been developed and is ready for
use against the following seven species:

(1)   Tsetse Fly (several species of Glossina)
(2)   Stable Fly (Stomoxys calcitrans)
(3)   House Fly (Musca domestica)
(4)   Caribbean Fruit Fly (Anastrepha suspensa)
(5)   Oriental Fruit Fly (Dacus dorsalis)

(6)   Gypsy Moth (Lymantria dispar)
(7)   Boll weevil (Anthonomus grandis)

There has been a great deal of research on other species,
however, for various reasons the technology has not yet been developed.
This includes mosquitoes, other Fruit Fly species, other Lepidoptera and
a number of other Diptera.

4.   Tsetse Flies

The development of the SIT for use against tsetse flies started
some 20 years ago. The progress made in spite of very limited resources
has been remarkable. Three aspects stand out as major advances. These
are (1) the great improvements in mass-rearing of the several species of
tsetse flies, including mass-rearing in Africa utilizing a locally
collected blood, (2) the availability of physical and chemical
attractants necessary for survey purposes and for use as attractant
devices to greatly reduce the wild population and (3) the ability to
assess the sterility induced in the wild population by the release of
sterile males.

The objective of using the SIT in an integrated programme to
eradicate one or more species of tsetse flies in a given area is to
arrive at a technology which is predictably effective at a minimum cost.
One should keep in mind that all of the tsetse eradication projects to
date which have utilized the SIT have essentially been research and
development projects. We have only now reached the stage where
reasonably accurate cost figures can be estimated and even these will be
considerably reduced in the near future as new developments in mass-
rearing systems and handling of sterile tsetse flies are further
developed.

In a tsetse eradication project utilizing the SIT the following
major events normally occur:

4.1.   Tsetse Survey and Population Suppression

Preliminary surveys are made to determine the species present in
their approximate density. At the same time pre-release populations
methods are either developed or adapted for use in a particular area.
Simultaneously mass-rearing facilities and methods are either developed
or adapted for local conditions.

The next phase in the operations include pre-release population
reduction followed by the release of sterile male tsetse flies.

During all of the above activities monitoring of wild and sterile
populations take place at regular intervals.

4.2.   Rearing

The use of the membrane or in vitro rearing, developed at
Seibersdorf, has greatly increased the efficiency of mass-rearing tsetse
flies. The use of larger rearing cages containing more mated females
will streamline the operation and reduce costs. The availability of
methods to eliminate contamination in blood by trypanosomes or bacteria
have permitted the use of locally collected blood and further reduced
tsetse rearing cost.

4.3. Fly Release

Although theoretically ratios of 3 sterile to 1 native male should be sufficient to eliminate a native tsetse population, the results of the Nigerian project indicate that a 9 : 1 ratio is required for effective and relatively rapid collapse of the native population. In addition, the use of techniques developed at our Seibersdorf Laboratory permit project personnel to follow the sterility induced into the wild population very accurately and thus, continuously adjust the numbers of sterile males needed for release in any specific area.

4.4. Eradication and Quarantine

Confirmation of tsetse eradication is usually done by combination of negative fly catches, release of sterile females which are re-captured and dissected to see whether they have mated with fertile males, and the use of sentinel animals to prove interruption of trypanosome transmission.

The final step in the programme is the establishment of quarantine barriers to prevent re-invasion into the area cleared of tsetse fly.

Although frequently omitted, plans for economic analysis of the project, surveys of trypanosomiasis incidence in livestock before, during and after the project and public relations are a vital part of any tsetse eradication project including those which utilize the Sterile Insect Technique.

4.5. Ongoing Field Programmes

Two of the current programmes (Burkina Faso and Nigeria) have been described during this meeting. Both of these programmes have resulted in the successful eradication of one or more species of tsetse fly in fairly sizable areas. However, it should be pointed out that both of these programmes were essentially research and development projects. The advances made by the expatriate and local staffs of these two projects and the practical application of the SIT against tsetse flies in West Africa has been remarkable. It can now be safely stated that the technology has been completely proven (something which should never have been questionned since the SIT had been proven to be effective against a number of other Diptera in practical programmes and had been demonstrated repeatedly against many other diptera, including tsetse flies, in research projects), that mass rearing of tsetse fly can be done in Africa utilizing in vitro technology, that the insecticide impregnated attractant devices are a significant improvement in population reduction prior to the release of sterile males, and that it is now possible to predict accurately, at least in the case of four species or sub-species of tsetse flies, that eradication can be rapidly and efficiently achieved utilizing a system which integrates one or more methods of pre-release population suppression with the release of sterile males.

One of the concerns of individuals involved with SIT projects for tsetse eradication has been whether the released sterile males would transmit trypanosomiasis. Results of detailed studies in Nigeria on G. palpalis palpalis have indicated that the procedure used prevents this from happening. The procedure is to feed the males once or twice in the laboratory before release. This plus the somewhat reduced life expectancy of the released sterile males prevents them from developing mature infections and transmitting the disease after release.

5.   Future

        The future utilization of the SIT in tsetse eradication projects will probably depend to a great extent on economics since efficacy and operations in Africa have been demonstrated. Some of the improvements which will reduce costs and increase efficacy are discussed below. The Joint FAO/IAEA Division will encourage R&D in these areas.

5.1.   Manufacturing

        Mass-rearing of tsetse flies is a manufacturing process very similar to the production of an insecticide. The product is the sterile male tsetse fly which is used in eradication projects.

5.1.1. Mass-rearing

        Great improvements have been made in mass-rearing during the past ten years and it is certainly realistic to assume that additional improvements will be made in the future. One of the reasons for this is that tsetse rearing methods continue to be investigated in many European laboratories where field studies cannot be conducted. Thus, any research laboratory in Europe which is involved in tsetse fly SIT projects rearing spend time on improving rearing methodology. An artificial diet which utilizes inexpensive forms of protein, carbohydrates, and fat should be possible. Although obligatory blood feeding insects have not been the subject of intensive studies to find alternative sources of nutrients, there is no reason to believe that it cannot be accomplished. Enormous progress has been made in reducing costs of rearing phytophagous insects through cheaper diets. Handling systems for tsetse fly mass-rearing have only been briefly investigated. Most of the specific techniques used are simple expansion of small scale rearing. Admittedly, the artificial membrane and progress in utilizing large rearing cages is a step in the right direction. However, if one looks at the mass-rearing of other insect species, it is quite apparent that significant changes must be made in tsetse rearing before it can realistically be termed mass-rearing. An inexpensive replacement for ATP is needed.

5.1.2. Quality Control

        With some insect species extensive work has been done to estimate quality of the product. This has been done to a limited extent with tsetse flies, and as mass-rearing is changed and improved more attention must be paid to the quality of the product being produced from the factory.

5.1.3. Mass-rearing Centres

        Regional centres to rear large numbers of tsetse flies for use in field programmes are a distinct possibility. However, one of the limiting factors in applying this concept is our inability at present, to separate the male from the female in the pupal stage. If male puparia could be separated from female puparia two to three weeks before adult eclosion, it would be much more realistic to consider regional tsetse mass-rearing facilities.

5.1.4. Transport

The transport of insects whether for SIT programmes, inundative release of parasites or predators or other reasons is a difficult undertaking. Only limited work has been done on the tsetse fly and much more will be required. Metabolic heat produced by pupae, water lost during pupation, vibration during transport, and external heat during transport, are all very serious deterrants to guaranteed delivery of a high quality product.

5.1.5. Shelf Life

One of the inherent problems in all SIT programmes is the rather short shelf life for the product. With insecticides and even microbial insecticides, the shelf life can normally be measured in years, whereas with the insects reared for SIT programmes, the shelf life is normally measured in hours or days. Additional work needs to be done on the shelf life of tsetse fly puparia to determine whether the puparial life, (currently approximately 30 days), can be extended signficantly.

5.1.6. Genetic Engineering

The explosion of theoretical and practical utilization of genetic engineering offers the possibility in the future, of genetically engineering the mother stock of tsetse flies so that at a certain temperature all males are produced whereas at a different temperature one gets more nearly a 1:1 ratio of males to females. There are certainly many other exciting potential uses of genetic engineering which may be forthcoming.

5.2. Use of the Product

Once the manufacturing plant has delivered to the customer the product (sterile male tsetse flies) it is up to the individuals involved to utilize this product most effectively. This will require rather extensive R&D. It is interesting to note that even today, after nearly 20 years of activity, research and development activities are still being undertaken to improve the use of endosulfan for aerial spraying. It is reasonable to expect that 20 years from now work will still be in progress on the most effective use of sterile male tsetse flies.

5.2.1. Release of Sterile Males

Release methods to date are primitive and will in the future need attention. For very large programmes, aerial release may be required for effective and faster distribution of the sterile males or prefered because of the lack of roads and trails needed for hand-releasing the sterile males. Aerial release technology in SIT programmes or parasite/predator inundative release programmes is still in its infancy.

5.2.2. Strategy

The strategic use of sterile male tsetse fly in an eradication

programme has received little attention. Until more SIT projects are
under way there is little likelihood that improvements in strategy can be
made, although theoretical strategies have often been prepared. These
must be evaluated under actual field conditions. The current strategy is
to release sterile males once or twice a week in a given area at distance
intervals which have been pre-determined by mark-release recapture
studies. Perhaps the use of an attractant device which has not been
treated with insecticides would improve contact between sterile males and
the wild population, which might increase mating between the sterile
males and the wild females. Can a surplus of females be utilized by
sterilizing them and releasing them into an area as a method of
population reduction? Serious thought needs to be given this idea and
well planned research and development undertaken on tsetse release
strategy in order to increase the efficacy and reduce costs of the SIT.

5.2.3. Pre-release Population Reduction Methods

        The recent development of the insecticide impregnated attractant
device is a major step forward in tsetse population reduction. This
method is inexpensive and effective against a number of tsetse fly
species. Since the method is relatively new, its full range of
possibilities and problems have yet to be determined. Some of our staff
in Nigeria have indicated the possibility of spraying blue paint on
trees, and wondered whether such an approach would not be as effective as
using blue cloth. Will an attractant device one fourth the size of that
presently used be effective? Are there more potent attractants, either
chemical or physical than those presently available? Anything that can
be done to reduce the cost and increase the efficacy of pre-release
population suppression will reduce overall costs of the project.

5.3.   Tsetse Behaviour

        An improved knowledge of tsetse behaviour followed by the
incorporation of this knowledge into SIT programmes will undoubtedly
increase the efficacy and decrease the costs of the SIT programmes. The
work in Zimbabwe on chemical attractants shows how good science will
result in major advances. Very simple methods coupled with very good
science can result in a major advance in our knowledge of tsetse
behaviour. There is a critical need for scientists to spend more time in
the field observing tsetse fly behaviour.

5.4.   Commercialization

        Commercialization of the entire SIT may not be feasible at
present, however, the commercialization of tsetse fly mass-rearing could
become a reality. Among the pre-requisistes for this taking place, are
(i) the development of a method of separating males and females at their
early puparial stage, (ii) safe and reliable methods of transporting
puparia and (iii) probably an increase in the shelf life of mass-reared
product. If the mass-rearing of tsetse fly were a commercial enterprise,
it would be a great deal easier to obtain comparative costs between the
SIT and other tsetse eradication technologies. It is not beyond the
realm of possibility that some bright business-oriented Africans will
decide to investigate the possibility of the commercialization of tsetse
mass-rearing.

Integrated sleeping sickness control*

P.de Raadt

Parasitic Diseases Programme, World Health Organization

Summary

Human trypanosomiasis occurs in the rural areas in 36 African countries. By force of circumstances the countries' health programmes' efforts have been reduced considerably over recent years. It is estimated that, as a result, amongst the 50 Million people at risk at present, only 5-10 Million are more or less protected by regular control programmes. The principles of control are systematic medical surveillance in combination, where possible, with local vector control. The paper reviews the various approaches and their respective advantages and shortcomings. Research over the last decades has resulted in simple new tools particularly for case detection and vector control. These methods being more suited for application by national health services with limited resources, do raise the hope that wider implementation of control measures can be envisaged in the near future. The participation by WHO and other international organizations in reviving national control programmes is discussed.

1. Introduction

Control of Sleeping Sickness is necessary in 36 African countries where the disease is endemic. It represents a heavy ·burden for the health services demanding precious medical facilties such as skilled personnel, transport, drugs and hospital beds and important demands on the resources of the vector control departments.

Today, due to economic restrictions and political upheavals, several health services have been obliged to reduce their Sleeping Sickness control programmes. This is clearly demonstrated from the medical surveillance results. Amongst the approximately 50 Million people exposed to risk, only 5-10 Million have remained under more or less regular surveillance. Consequently, severe epidemic outbreaks have occurred, for instance, in Sudan and Uganda, and, on a smaller scale, in Cameroun, Congo, Ivory Coast, Kenya and Zaïre. These alarming demonstrations of the potential danger of leaving endemic foci unattended created an increasing awareness at national and international levels and activated research towards better technical solutions for prevention and control. As an example, when WHO together with the World Bank and UNDP developed the Special Programme for Research and Training in Tropical Diseases, trypanosomiasis was selected as one of the six priority diseases.

* This paper is largely based on: "Sleeping Sickness Control: a new elan" by P. de Raadt, distributed in 1983 by the Biological Division, Smith Kline RIT and the Institut de Médecine Tropicale "Prince Léopold, Antwerp, Belgium.

To improve the impasse in control, the starting point was, that public health needed simpler control tools which could be used by less, or preferably, non-specialized personnel or that can be handled at community level.

Over recent years, research in the endemic areas has led to new knowledge on the transmission cycle and on the clinical aspects. It has also been remarkably productive in the development of new control tools such as the serodiagnostic tests appropriate for field use, a more sensitive method for parasite detection in blood, new drugs for treatment and simple vector control devices that can be applied by the rural communities themselves.

2. The Principles of control

For bringing a vector-borne and man-to-man-transmitted disease under control there are two options. One is to reduce the reservoir of the parasites by systematic detection of the infected individuals and treat them all on the assumption that, with the last infection cured, the parasite would have disappeared. The second option is to prevent transmission by reducing man's exposure to tsetse bites.

The principle of systematic surveillance and treatment of the population at risk was introduced in the thirties with the well-known campaigns based on the work by Jamot and his successors. None of these impressive, well-organized programmes, however, led to the eradication of the parasite. At the most, one could obtain at long-term, a prevalence level of 0.01% which remained constant in the well-surveyed areas. The persistence of the parasite was assumed to be due to the limited sensitivity of the diagnostic methods by which a residue of infected people remained undetected at surveillance. Labusquière spoke of the "zero épidémiologique": the minimum level or prevalence below which even the most intensive surveillance could not improve results any further. By the early fifties, this level was reached and maintained in most of the West and Central African countries. The introduction of serological tests such as the indirect fluorescent antibody technique some 15 years ago, improved the surveillance detection rate by 30-50%.

An effort to eradicate the small residual reservoir in man by pentamidine prophylaxis was made in the fifties in Sudan and followed in several other countries. Here again, though impressive results were obtained and the word "eradication" reappeared in several plans and programmes, in none of the foci total eradication was reached. The likely explanation was that either new parasites could have been reintroduced by infected people or flies or, as clinical observations suggested, that trypanosomes had become resistant to pentamidine. Another disadvantage by which pentamidine prophylaxis became in discredit was the occurrence of "camouflaged" infections amongst the people who had been treated whilst already infected. At present, it is still successfully practised in a few countries but it requires considerable extra expenses in addition to the costs of the surveillance teams which will have to be maintained whether prophylaxis is used or not.

However, recent epidemiological findings may provide another explanation why human trypanosomiasis persisted in practically all of the historical foci in spite of some 50 years of intensive control . According to the classical viewpoint, T. b. gambiense is a man-to-man transmitted parasite with its reservoir of infection in man and a small part of it in the tsetse flies. Recent discoveries in West Africa, however, showed that the wild and domestic animals were naturally infected with trypanosomes of the T. b. brucei species which biochemically proved identical to T. b. gambiense. It is impossible to judge at this moment to what extent these trypanosomes form a direct source of infections to man. Transmission from animals to man may be a rare event, but the possible existence of a residual reservoir of T. b. gambiense in animals is a new element in the epidemiology.

As for vector control, in the early days, handcatching and trapping were the only techniques available. These methods, if practised continuously, could lower the transmission rates fairly effectively. After the Second World War, insecticides were introduced. They were used in East Africa to control T. b. rhodesiense disease and on a large scale against animal trypanosomiasis, for example in Nigeria and more recently in Botswana and Zimbabwe. This type of campaign is costly and requires continued surveillance and respraying to maintain the treated areas free from flies. The risk of reinvasion has recently been demonstrated in field experiments which showed that the range of displacement by tsetse flies can reach as far as 20 km. There is no case for eradication even locally, unless the fly's habitat can be destroyed at long-term by modifications of the landscape

Since the movements of man are unpredictable and practically uncontrollable, one can not fully protect people in Africa from being bitten by tsetse flies. The local tsetse challenge, however, can be reduced by control measures in the heavily infected areas. A good example of the past is the Busoga area in Uganda where fly densities were maintained low during many years by regular systematic groundspraying. Aerial spraying as practised experimentally in the Lambwe Valley, Kenya (1) and in West Africa (2) and, recently, for control, in Uganda, can be used in certain situations. However, due to the cost, its application for Sleeping Sickness Control remains restricted to emergency operations.

The principles of Sleeping sickness Control have not changed over the last fifty years: the basis and first line of defence is regular surveillance of the population for the purpose of trypanosomiasis control as well as for preventing those who are infected from developing the late stage of the disease by which the risk of sequellae or treatment failures can be reduced. In addition to surveillance, wherever possible, local measures for fly control should be applied, particularly where the fly density is high, where surveillance results indicate an increasing prevalence or simply because flies may be a serious nuisance for the local population. It is also important to ensure that wherever cattle industry requires tsetse control for the prevention of animal trypanosomiasis, maximum advantage should be taken to include Sleeping Sickness control at the same time.

3. The new tools
For surveillance, the most promising new methods are the serological tests developed for field use such as the direct agglutination test on cards (3) and an indirect agglutination test adapted to field use (4). One important advantage of these serological tests for population screening is that, being simple, they permit involving large groups of health personnel in surveillance and offer a realistic means for improving population coverage by the national services. The second main advantage is that like the fluorescent antibody technique, they are more sensitive screening tools than the classical parasitological methods used in the field and may in the future allow for prolonging the intervals between the mobile teams' visits, as suggested by Van Nieuwenhove (5).

In areas where people are used to agricultural spraying, ground spraying can be practised cheaply and effectively by the villagers with low dose insecticide applications once or twice a year. Present revived interest in control by trapping is a most interesting development since traps and impregnated screens can be handled by the rural communities themselves. The very cheap and simple traps now designed after the models by Challier, Laveissière (6) and Lancien (7) are highly suitable for local control of Glossina of the palpalis and fuscipes groups and their use in West Africa and Congo showed that reductions of the local fly density of over 95% can be achieved (8). Pilot trials in Ivory Coast (9) and Congo showed that rural populations gain confidence in these techniques very rapidly and that they can be motivated to apply such tools.

4. National and international control strategies
    The trypanosomiasis control programmes are usually national activities except in some emergency situations. The investments required depend on the extent and severity of the Sleeping Sickness problem in the country and have to be weighed against other national health priorities and against the available resources, particularly skilled manpower.
    In most countries it will be necessary to choose a more or less satisfactory compromise between the ideal and the strict minimum quality level of control. The ideal level is systematic serological screening once a year of the entire population at risk,, subsequent parasitological examination of those who are found to be seropositive and hospital treatment of those who have been confirmed by parasitological examination. Control by surveillance and treatment should, where possible, be combined with local vector control either by groundspraying, traps, or impregnated screens or any combination of the three. At the other end of the scale, at the lowest level, is serological screening only, followed by treatment of all seropositives: no doubt a defendable, though not preferred compromise, but practised in certain countries. Between these two extremes many combinations are possible.
    The role of the primary health care worker will differ from country to country. They should always be involved in identifying suspects on the basis of symptoms like swollen glands, or fever or on people's history and refer such suspected patients to the health centres for further examination. They may also collect blood slides or serum samples for diagnosis elsewhere. A very important role for the primary health care worker would also be to organize the application and maintenance of local vector control devices like traps and screens. They can also participate in the follow up on patients after treatment and ensure their regular follow-up examination at the health centres.
    For wide-scale use of the field serological tests, traps and screens, much will depend on the costs which at present are above the average of what national budgets can afford for Sleeping Sickness control. Bilateral or other international support will be needed for the supplies of the majority of the countries during the following years. It should be noted that the recent technical developments have inspired new confidence in the feasibility of expanding or reviving national control programmes and some countries have responded by increasing their national budgets for Sleeping Sickness Control.
    The same sense of renewed confidence was reflected in resolutions adopted by the WHO Regional Committee in 1982 and the World Health

Assembly in 1983 requesting coordination and technical support, training and fund-raising for implementing active participatioon in simple control methods.

The role of WHO and other international organizations will be (1) to coordinate control between the countries and to participate in the national planning of the control programmes; (2) to provide training at national as well as international level and (3) to act as intermediary in supplies and in mobilizing funds where necessary.

Provided mobilization of international resources remains successful, and national and international motivation continues to be as strong as it is now, one may expect a significant improvement in Sleeping Sickness Control in the near future.

REFERENCES

1. BALDRY, D.A.T. (1972). A history of Rhodesian sleeping sickness in the Lambwe Valley. Bull. Wld Hlth Org. V 47, 699–718

2. MOLYNEUX, D.H., BALDRY, D.A.T., DE RAADT, P., LEE, C.W. and HAMON, J. (1978). Helicopter application of insecticides for the control of African Human Trypanosmiasis in the moist savannah zones. Ann. Soc. belge Méd. trop. Vol. 58, 185–203

3. MAGNUS, E., VERVOORT, T. & VAN MEIRVENNE, N. (1978). A card-agglutination test with stained trypanosomes (C.A.T.T.) for the serological diagnosis of T. b. gambiense trypanosomiasis. Ann. Soc. belge Méd. trop., 58, 169–176

4. BONE, G.J., CHARLIER, J., (1975). L'hémagglutination indirecte en capillaire: un méthode de diagnostic de la trypanosomiase applicable sur le terrain. Ann. Soc. belge Méd. trop., 55, 559–569

5. VAN NIEUWENHOVE, S. (1984). personal communication

6. CHALLIER, A. & LAVEISSIERE, C. (1973). A new trap for capturing glossina (Diptera Muscidae): description and field triaés. Cahiers ORSTOM, série Entomologie médicale et Parasitologie, 11, 251–262

7. LANCIEN, J. (1981) Description di piège monoconique utilisé pour l'élimination des glossines en République Populaire du Congo. Cahiers ORSTOM, série Entomologie médicale et Parasitologie, XIX (4), 235–238

8. LAVEISSIERE, C. & COURET, D. (1980). Traps impregnated with insecticide for the control of riverine tsetse flies. Transactions of the Royal Society of Tropical Medicine and Hygiene, 74 (2), 264–265

9. LAVEISSIERE, C., HERVOUET, J.-P., MEROUZE, F. & CATTAND, P. La campagne pilote de lutte contre la Trypanosomiase humaine dans le foyer de Vavoua (Côte d'Ivoire): Bilan de la campagne: les prospections médicales et la participation de la population. Cahiers ORSTOM, série Entomologie médicale et Parasitologie, in press.

Integrated tsetse control: Present and future programmes of national and international organisations

K.B.David-West

Federal Livestock Department, Lagos, Nigeria

SUMMARY

Tsetse infestation in Africa constitutes a major constraint to live-stock development, as seven million sq. km. of potential grazing land is made inaccessible to livestock for grazing. In Nigeria about 75% of the total land area is tsetse infested. A tsetse eradication campaign launched about thirty years ago have cleared about 250.000 sq.km of land, which is now used for agriculture and other purposes. The method of control is by ground and aerial insecticide application. In 1979 a pilot integrated Sterile Insect Technique programme was started and has now cleared about 1000 sq.km of land. A second phase of the programme is planned to cover 10.000 sq.km.

A lot of work has been done in the field of tsetse control by Donor Agencies and national Governments but there is a general lack of coordination of these efforts. There is lack of an integrated control programme for Africa. Such a programme is necessary and should be implemented as a collective action between Donor Agencies and National Governments.

Tsetse infestation in Africa constitutes a major constraint to live-stock development, as seven million sq.km. of potential grazing land is made inaccessible to livestock for grazing purposes. The overall loss of beef output is about 1.5 million tons annually. According to Lambrecht (1964), trypanosomiasis has determined which areas of Africa is habit-able by man and his livestock and has made the African the main source of domestic labour since it prevents the extensive use of work oxen.

In view of the increasing human population in Africa and the dete-riorating protein output in the continent, the FAO in 1974 prepared a food plan for Africa in which there was a considerable focus on a joint international tsetse eradication programme. It is a matter of regret that inspite of appreciable efforts of both National Governments and Inter-national organisations the tsetse and animal protein problem still re-mains unsolved. The apparent lack of success is primarily due to lack of coordinated efforts for a concerted continental tsetse eradication pro-gramme as indicated in the FAO Food plan for Africa. Secondly most of the national governments lack the resources, particularly manpower and the foreign exchange to procure the necessary inputs for implementation of tsetse control programmes. Some governments could not also afford the cost of incentives necessary to motivate field workers, who spend pro-tracted and tedious working days in remote rural areas.

Tsetse control in Nigeria

The land area of the country is about 941,849 sq.km. out of which 75% is still tsetse infested. Tsetse eradication campaign embarked upon by the Government of Nigeria since thirty years ago, have cleared about 250.000 sq.km. of land. Although there is no current statistics on the incidence of trypanosomiasis in the country, field reports indicate wide-spread sporadic cases of the disease in cattle, sheep and goats. The situation is worsened by seasonal drought, which causes over-grazing, land degradation and high livestock mortality.

The methods of tsetse control include chemical methods using ground and aerial spraying with insecticides. In view of the public outcry against environmental pollution due to insecticide application and the difficulties of control methods adapted to the Southern Guinea and forest zones of the country, the Nigerian Government started to explore alternative non-pollutant control methods. In 1979 a pilot biological control programme was started at Vom as a joint project between the Nigerian Government and the International Atomic Energy Agency.

Achievements of the Programme are :
(1) More than 100.000 producing females are now being maintained in Vom Laboratory. The females produce averagely 10.000 pupae per week and these provide 4.000 sterile males weekly field releases.
(2) 60% of the 1.500 sq.km. target area in Lafia LGA of Plateau State has been cleared of G. palpalis. Total eradication of the area is expected by early rains of 1986.
(3) Target area is effectively protected against rainfestation by the use of insecticide impregnated screens.

A second phase of the project covering an area of 10.000 sq.km. initiated in the Plateau State of the country, which is suitable for the rearing of high grade animals. Although the Sterile Insect Technique (S.I.T.) is apparently successful, the initial investment costs for establishment of infrastructure (insectories and equipment) and technology transfer are relatively high. Its application over large areas as in Nigeria will require either the establishment of several insectories or the use of aircraft for sterile insect distribution to remote control areas. Cost-saving research support and strong training component will be required for successful field application of the S.I.T. programme.

The tsetse control policy includes :
(1) Continuation of tsetse eradication programmes by integration of feasible methods viz, ground spraying, aerial spraying, use of traps and impregnated screens and S.I.T. methods.
(2) Utilisation of grazing areas with low and medium tsetse challenge supported with chemotherapy.
(3) Utilisation of trypanotolerant cattle in tsetse infested areas.

Issues for consideration in the implementation of Tsetse Control Programmes
1. Tsetse distribution is dynamic and there is need for periodic tsetse distribution surveys.

2. The actual incidence of trypanosomiasis, loss of production and mortality due to trypanosomiasis is not quantified in several African countries. These facts are required for a proper assessment of the tsetse problem and an economic evaluation of the control programmes.

3. The data on range resources and animal productivity prior to and after tsetse eradication are vital for planning and evaluation of tsetse programmes.

4. The problem of drug resistance and the production of newer, more effective drugs needs to be considered. The search for a vaccine to control trypanosomiasis deserves urgent attention.

5. Donor Agencies have not given sufficient support to chemotherapeutic control of trypanosomiasis. Research into drug resistance, vaccine and new, less toxic, and effective drugs deserves priority attention.

6. The monitoring of insecticidal effect on fauna and flora is essential to assess periodically, the pollution effects on the environment.

7. New control methods particularly the use of traps and impregnated screens have been exciting. However they still need further perfection. The search should continue to find less toxic insecticides and other cost-effective control measures which village communities can implement themselves.

8. Consideration should be given to the appropriate balance between upstream and downstream research that can solve practical problems under field conditions. Considerable research work on tsetse is being implemented outside Africa. Collaborations with African Research Institutes is essential and their research capabilities need further strenghtening.

9. There are weak information linkages between Anglo-phone and franco-phone Africa. This can be improved by encouraging exchange visits and strenghtening the FAO tsetse bulletin.

10. Coordination of regional and sub-regional projects have been lacking. Donor Agencies seem to be operating in isolation of each other. Although the FAO has set up an African trypanosomiasis Commission and related panels, there is little impact on field control programmes. There is yet no plan for a coordinated regional or sub-regional control programmes and a targeted time frame-work for tsetse eradication on the continent. The FAO has established the 1st sub-regional development support unit for West Africa but this programme cannot offer practical solutions to the tsetse problems within the subregion since several contigious countries including Nigeria are excluded from the programme.

For the future, we need to improve the data base for African Tsetse and Trypanosomiasis, and elaborate a master control programme within a time framework and funding schedule as a collective action between National Governments and Donor Agencies. The programme should include research, training, development and integrated land use.

Integrated control of Glossines: Action of the Commission of the European Communities

C.Uzureau & M.De Bruycker
Directorate-General Science, Research and Development, CEC

The first Science and Technolgy Programme for Development was adopted by the Council of the European Communities on 3 December 1982 (OJ of 17 May 1983) and covered the period 1983 to 1986.

Implemented by DG XII (Science and technology for development and cooperation with developing countries) in cooperation with DG VIII (Development), the programme aims at strengthening and promoting scientific and technical activities on the part of Member States and developing countries. It comprises two sub-programmes. "Tropical Agricultural" has a budget of 30 million ECU over 4 years, three quarters of the total programme funding, while "Medicine, Health and Nutrition in Tropical Areas" has a budget of 10 million ECU over the same period.

In the sub-programme "Tropical Agriculture", the term "agriculture" is taken in it widest sense and therefore includes the following fields:
- the improvement of both food crops and crops for processing, although food production will remain a priority;
- crop protection including integrated control of insects and weeds;
- post-harvests technology;
- animal husbandry, including veterinary medicine (stressing parasitosis) and fodder crops;
- fishing, fish-farming;
- forestry;
- water resources;
- problems of erosion, fertility and soil-regeneration.

The Commission, aided by a Management and Coordination Committee, has selected more than 220 projects out of 1300 proposals from 73 countries including 9 Member States of the European Community.

Some 42 to these projects concern animal husbandry and, in particular, veterinary medicine. Some involve the study tripanosome-resistance in cattle and sheep, how this is genetically determined and the interactions between medication and immune-responses to African tripanosomiasis. Another approach to tsetse control involves research into pheromones and attractants to enhance the effectiveness of traps.

The procedures used in the above sub-programme were also applied in the sub-programme "Medecine, Health and Nutrition in Tropical Areas". 168 projects were selected from 660 proposals submitted by 65 countries, of which 9 were Member States and 56 developing countries.

183 contracts were then signed, covering the following fields:
- parasitic diseases: e.g. malaria, schistosomiasis, filariasis, African tripanosomiasis, Chagas' disease, leishmaniasis;

- microbiology (including African aspects of AIDS);
- genetics, including haemoglobin disorders;
- public health including the health of mother and child;
- nutrition.

14 contractors concerned research on African tripanosomiasis, with the stress falling mainly on antigen variation (the major obstacle to the development of a vaccine), using molecular-biology techniques, and on research into the development of new medication. Several research teams are studying molecular and genetic aspects and their application in tripanosomiasis control. Preliminary results in this field are very promising: e.g. the discovery of the haploidy of metacyclic tripanosome forms and the clarification of the difference between the glycolytic enzymes of tripanosomes and those of their mammalian hosts. This opens up new potential for the development of vaccines and medication. Field research includes epidemiology and vector (tsetse fly) control. Four contracts concern research into tsetse-fly traps, tripanosome-resistance and resistance to chemotherapy.

At this stage of the programme, a number of conclusions can already be drawn in relation to its impact:
- it has been possible in certain cases to slow down the fall-off in Europe's potential for tropical research.
- the research potential of developing countries also seems to have benefited from this programme since 112 contracts are in relation to proposals from these countries.
- North-South and South-South cooperation has been strengthened.

A second Science and Technology Programme is being prepared and will be submitted to the European Parliament and to the Council of Ministers in 1986, with a view to implementation, after a first call for proposals, from 1987 onwards.

These proposals must concern research as part of realistic projects capable of practical application in the short or medium term. They should also stimulate or consolidate links or agreements between research teams in Europe and those in developing countries, leading to the eventual development of research networks or, at least, to more efficient exchange of information.

Tripanosomiasis and the technology for combatting the disease and its vectors will remain one of the priorities of the programme.

This symposium could therefore serve to encourage specialists, whether present or not, to undertake joint discussions and draw up proposals for submission in due course, to the European Commission or other funding agencies.

Session 4
General aspects on Glossina

Chairman: C.Pelerents

Tsetse fly control and primary health care

R.C.Pinhão
Disciplina de Entomologia, Instituto de Higiene e Medicina Tropical, Lisboa, Portugal

Summary

The author considers that the present possibilities and ressources of tse-tse fly control are compatible with its integration on PHC activities. Some of the main aspects to be considered before the integration is achieved are presented and briefly discussed. They relates with two great areas: critical review of present scientific knowledge and local ressources, and problems connected with the participation of the populations involved.

Primary Health Care (PHC) were defined in 1978, at the Alma-Ata Conference as the foundamental strategy in community medicine. Since then, the evolution of the public health services of a vast majority of countries have been re-markable, specially in the so-called Third World. We may say that PHC concept is today firmly established with a world wide acceptance.

Probably, economic reasons are the more important argument for its implementation. PHC is the sanitary strategy with a more favorable costs-benefits ratio. Constrained, as they are, by tremendous economic problems, but well aware of the importance of the health of their peoples, the Third World Governments have invested in PHC area most of their health ressources.

International health authorities also have heavily invested in the promotion of PHC and WHO, as you well know, have excelled in this orientation.

Being practically the only sanitary structure firmly established, PHC are today the more obvious support for vector born diseases control and therefore for vector control operations. This fact and the financial and logistic restreints that are the rule inAfrica, makes the integration of V.C. operations in PHC the more realistic attitude.

But this integration raises several problems and difficulties and unless they are correctly solved the integration may not work at all. I think is worth while to bring here some considerations on that matter.

Underlying the PHC definition, are several foundamental concepts and the more important are:

- its organization in a communitary base
- the outstanding and essential rôle assumed by the participation of the population
- its multidisciplinar and intersectorial character
- its integration with the national economic and social development programs

By population participation is intent notthe mere acceptance or collaboration, but its active engagement at all steps, since planning to execution and control.

It is obvious that the transition of vector control operations from independent and autonomous units to the PHC system imposes profound changes in strategy and mentalities predominant in those units and technicians. Specialists have to adapt themselves to the idea that its rôle is only to propose, to advise, to teach, to perform, to control but not to decide. Great projects, limited in the space and on the time probably will no longer be possible, g iving place to less spectacular programmes, less ambitious but more realistic. The big monodisciplinary teams will be fragmented and its technicians reeducated for multidisciplinary activities. The philosophy of "the best possible results" must be replaced by an old saying of my country: "best is good's enemy".

But, for the reasons presented, integration of vector control operations in PHC is a need and I have no doubts that it will be done with everybody cooperation. The problem is not if, but how.

In what concerns tse-tse fly control a previous question must be answered. Are the control methods presently available compatible with the integration in PHC? The answer can only be affirmative. Both in case findin g and vector control we dispose presently of techniques that are well adapted to be implemented with the participation of the population. CATT and m-AECT in the first case and traps and screens in the second are techniques particulary fitted to be manipulated at community levels.

But if integration can be fore-seen as possible and promising the way it must be done is not yet clear. The integration of tse-tse control in the PHC depends of the answer to a certain number of questions before the planning phase can be initiated. Many of these questions are of local significance but a general orientation may be tried. Among the more important I think that the following two groups of questions are of paramount importance.

1. Critical review of the present knowledge and ressources

1.1 Objectives (erradication? interruption of the transmission? reduction of fly-man contact?)

1.2 Techniques and means (including the "non-entomological" means of reducing the fly-man contact, as water and sanitary facilities)

1.3 Sanitary occupation

1.4 Ressources available (financing, equipment, staff, etc.)

1.5 National health and socio-economic objectives

2. Community participation

2.1 Degree of adaption of the V.C. operations to the habits and activities of the population

2.2 Degree of dependence of the V.C. programs from the community ressources

2.3 Degree of expected colaboration of the population in V.C. activities

2.4 Activities that can rely entirely in the population participants

2.5 How can the V.C. operations be conjugated with the others activities of the population

2.6 How to connect the community participation in V.C. programs with others ac tivities of the PHC

2.7 How can the participation of population be stimulated and improved?

There are some of the more important points to be analyzed before integration is decided and planning begins. The perfect understanding of the philosophy of both parts involved - community medicine and vector control - is of decisive importance for the success of any control program.

This subject, by itself, needs a very large change of opinions from specialists of both fields and justifies a new meeting like this one. I leave here the idea to the consideration of the organizers of this meeting.

Tsetse research and development at Seibersdorf: A review

A.M.V.Van der Vloedt, D.Luger, J.P.Kabayo & E.D.Offori
Entomology Unit, FAO/IAEA Agricultural Biotechnology Laboratory, Vienna, Austria

Summary

Tsetse research at the Seibersdorf Laboratory was initiated in 1968. At
that time, revolutionary events generated increasing interest in the use
of non-pesticide methods for vector control, namely, (1) eradication of
the screwworm fly in the United States of America by means of the sterile
male release technique, and (2) success in maintaining self-supporting
colonies of economically important tsetse species in laboratories at
Lisbon, Bristol and Paris.
Against the background of well documented but sometimes underestimated
biological and technical problems associated with the use of living
animals and artificial feeding systems for rearing an obligatory
haematophagous insect of low reproductive potential, research efforts were
directed toward development of large-scale rearing and radiation
sterilization of target species of tsetse flies.
Co-operative research efforts and active exchange of information between
the Seibersdorf staff, consultants and scientists participating in the
Joint FAO/IAEA co-ordinated research programme have made it possible to
determine realistic production standards and to re-assess factors
affecting the efficiency of the sterile insect release method for tsetse
fly control or eradication.
A continuous sequence of technical improvements enhanced establishment at
the Seibersdorf Laboratory of a membrane-fed stock colony of <u>Glossina
palpalis palpalis</u>, containing 75,000 breeding female flies, with a monthly
production of 150,000 puparia and a distributable excess of at least
60,000 puparia per month. Moreover, other species (<u>G. tachiniodes</u>, <u>G.
pallidipes</u>) have been successfully adapted to the <u>in vitro</u> system.
There have been recent advances in the continuing search for further
refinement of the technology of mass-rearing of tsetse, such as the
development of synthetic diets, emergence and larviposition
synchronization and the use of large holding cages.

1. Introduction

At the inaugural session of the 1st Symposium on Tsetse Fly Breeding under
Laboratory Conditions and its Practical Application, convened in Lisbon, in
1969, Dr. T.A.M. Nash of the Langford Tsetse Research Laboratory declared:
"Our ability to rear <u>Glossina</u> successfully has pushed the tsetse and
trypanosomiasis problems out of Africa into the world's centres of learning
and research".
Indeed, successes in tsetse colonization achieved in Lisbon, Bristol and
Paris during the 1960s announced a new era in the study of tsetse fly biology
and trypanosome transmission. Moreover, they enhanced toxicity studies of new
insecticides and the exploration of alternative avenues of tsetse control [1].

Meanwhile, the mounting interest in the use of non-pesticide methods of insect control and the successful implementation of the sterile male technique against the screwworm fly in the United States, had encouraged speculations regarding potential possibilities of the sterility method for tsetse control and eradication [2].

Scientists notably T.A.M. Nash, J. Fraga de Azevedo, J. Itard and their associates, who had been successful in transferring wild material from various parts of Africa to their laboratories in Europe were members of a panel of experts who, in 1971, recommended that the Seibersdorf Laboratory be strengthened to support the Sterile Insect Technique (SIT) projects on tsetse flies. The panel also recommended that studies should be done to develop methods for artificial feeding and sterilizing the major vector species of Glossina [3].

2. Laboratory colonisation of important tsetse species

Tsetse research at Seibersdorf was started in 1968 with G. morsitans morsitans (Rhodesia strain) and G. austeni (Zanzibar strain) material derived from the Lisbon and Langford colonies respectively.

G. tachinoides (Chad strain) and G. p. palpalis (Zaire and Nigeria strains) were introduced in 1972 and 1974, taking advantage of existing colonies in Maisons-Alfort, Paris and Antwerp, Belgium. A Tanzania strain of G. m. morsitans was built up from field-collected puparia in 1973.

More recently, experimental colonies of G. pallidipes (Uganda and Kenya strains) were established as offshoots from the Amsterdam (1978) and ICIPE'S Mbita Point Field Station (1984) colonies.

During the first few years, attempts to maintain the above mentioned species were based on methods consistent with existing practice, e.g. feeding the flies on rabbits ears, flanks of goats and guinea pigs. These methods were used at the laboratories from which the insects were obtained [4]. Later a variety of modifications and refinements were introduced.

From the initial rearing trials to the present, the trends of research and development at Seibersdorf can be divided into the following categories: (1) reproductive physiology, (2) rearing on live animals, (3) rearing in the absence of living hosts, including the development of chemically defined diets (4) radiation sterilization.

2.1 In vivo and in vitro rearing

The availability of laboratory adapted strains made it possible to study their metabolic and reproductive efficiency and to assess their competitiveness with natural populations [5]. In addition, the factors concerned with laboratory adaptation of tsetse flies and their nutritional requirements were systematically investigated with special emphasis on the development of artificial feeding techniques. The latter required a workable prototype of "surrogate host" consisting of a grooved glass plate into which suitable types of uncoagulated blood were placed, covered by an aseptic agar/parafilm membrane and the entire unit maintained at 37-38°C by a heating device [6].

Inspite of stringent precautions against contamination during blood collection and selective approaches for blood preservation, usually the reproductive performance of female flies fed entirely on the in vitro regime (6 days a week) deteriorated from generation to generation. Thus, the continuity of existing colonies could only be ensured by boosting the flies' blood intake with an in vivo supplement on rabbit.

Problems associated with the use of animal skins and agar-agar membranes, the laborious procedure for preparing the agar/parafilm feeding unit and the vulnerability to break-down of the circulating warm-water system, made it necessary to look for more convenient systems. Of the many improvements that have made progress possible, the development of silicone membranes simulating the host skin and the design of electric heating plates for warming the blood membranes were perhaps the most significant [7][8]. These technical innovations together with special care in the preparation of materials for feeding, were largely responsible for the successful maintenance of a productive all-membrane-fed self-supporting colony of the Tanzania strain of G. m. morsitans using defibrinated equine blood. This colony, kept as a back-up for the Tanga Tsetse Research Project, comprised 10,000 female flies at the end of 1977 and was for some time a reliable source of excess material for distribution.

As the emphasis from early on was toward developing in vitro rearing, there was little effort to make major changes in the in vivo practices. When small numbers of flies were needed e.g. to support research studies, or the membrane-fed flies in the then existing G. m. morsitans and G. tachinoides colonies needed an animal supplement, rabbits were adequate hosts [9]. However, with the introduction of a guinea pig-adapted strain of G. p. palpalis during 1974, -in anticipation of a SIT programme in Nigeria-, it was necessary to establish an animal-fed colony as a source of material for future experimental in vitro work. Thus, at the time the Nigerian project, BICOT, became operational in 1979, a highly efficient in vivo system was available [10].

During the period 1976-1982, further developmental work at Seibersdorf and at the Langford Tsetse Research Laboratory led to a breakthrough in in vitro technology. Procedures for using different types of blood and criteria for determining their suitability were established. Interestingly, the superiority of whole pig blood to cow blood was demonstrated for most species of tsetse [11][12] together with some inherent risks associated with its use. It was repeatedly experienced that bacterial contamination, initiated through improper blood collection at the slaughter-house or by feeding colony flies with contaminated equipment caused high mortality as well as the less obvious pathological signs such as reduced fecundity and small offspring. These deterrent effects were greatly reduced by aseptic precautions for all steps in diet preparation and routine high dose radiation (50 to 100 krad*) for final decontamination of the diet before feeding [13][14]. Recent studies also have shown that some antibiotics (e.g. chloramphenicol, polymyxin B sulphate combined with bacitracin) can be used therapeutically to eliminate residual contamination in blood and eventually from mature flies without harming the flies (Gingrich and Van der Vloedt, unpublished data).

Another step toward optimization of the membrane feeding technology for large-scale tsetse rearing was the follow-up on an earlier idea [11], that lyophilisation of whole blood might offer a more practical solution to the problem of storage and transportation. When it was found that flies did not require intact cells and would feed on haemolysed blood, the way was open to develop a suitable freeze-dried product, which after being reconstituted and supplemented with the phagostimulant ATP, constitutes an adequate diet [15][16].

Thus, the Seibersdorf Laboratory was able to meet one of its major original goals, namely, to develop an in vitro system for mass-rearing tsetse flies that eliminates expenditures on the maintenance of living host animals.

* 1 rad = 1.00×10^{-2} Gy

The success in meeting this goal is demonstrated with the maintenance of a large back-up colony of G. p. palpalis in support of the on-going BICOT project in Nigeria. (Tables I, II and Figs. 1 and 2)

The data on survival, fecundity and quality of the offspring gathered for the G. p. palpalis production units, in which female flies are kept for a life span of 80 to 85 days, indicate a performance which exceeds all previously achieved values of innate capacity for increase in numbers [12][17].

An important contribution to the use of blood collected at the slaughter-house, was the development of a very comprehensive and sensitive quality control test which is used to screen diets before use to ensure their complete suitability for feeding (Van der Vloedt et al., in preparation). This bioassay, in which all possible biological parameters, e.g. survival, fecundity, offspring size and pregnancy dynamics during a 25-day period after female eclosion, are given a numerical score which is indicative of the blood quality, has enabled us to avoid deleterious effects of drugs used in the nutrition and medication of blood donor animals. Results of quality control tests on several hundreds of samples of both fresh and freeze-dried porcine and bovine blood, confirmed the previously known nutritional superiority of pig blood over cow blood and also stressed the need for optimizing the reconstitution of freeze-dried blood in general. In this respect the beneficial effect of sonication on particle size in resuspended blood and consequently on fly performance was clearly demonstrated [18][19]. Moreover, results of recent studies on the biochemical basis of differences in nutritional qualities between pig and cow blood [20], give evidence for the suitability of the diet containing equal volumes of processed fresh bovine blood and reconstituted freeze-dried or processed fresh porcine blood, used since 1982 for rearing G. p. palpalis.

2.2 Research on synthetic diet

The successful development of an in vitro feeding technique using fresh or freeze-dried blood fed through a membrane instead of live hosts has introduced economic and logistic advantages in the rearing of tsetse flies. In spite of these advantages, however, in vitro feeding of flies on whole blood has certain limitations: a) it demands stringent aseptic measures in the collection, storage and handling of blood, b) blood is subject to contamination by drugs, etc. used in the nutrition and veterinary care of donor animals known or thought to have deleterious effects on tsetse, c) there are many differences in the composition of blood induced by a variety of causes (including genetic, nutritional, physiological, stress and other factors) some of which could influence the suitability of a particular batch of blood for feeding to tsetse.

There was therefore need to find a replacement/an alternative to blood as a diet for tsetse. During the last few years research in Seibersdorf has been aimed at developing a diet whose nutritional value could be standardised and which could be free from contaminants harmful to tsetse. A semi-defined synthetic diet composed of commercially available ingredients has been developed and used successfully to rear G. p. palpalis and other species of tsetse [21]. Further work is in progress to reduce the costs and molecular complexity of the ingredients without sacrificing the nutritional quality. Collaborative research work in which the diet is used in studies on trypanosome infection and maturation in the vector have been initiated.

2.3 Radiation sterilization of tsetse flies with gamma rays

Since 1968-1969 studies have been conducted to determine for several species of Glossina the irradiation dose which gives the optimal combination of male survival, sexual competitiveness and steriliztion. Dose/response relationship for radiation doses between 1 and 30 krad (1 krad = 10 Gy) were established during survival and fertility tests with both young adult male flies and males treated as late puparia.

Work in Maisons Alfort, Paris [22], Langford, Bristol [23][24] and Seibersdorf [25] and during the pilot field projects in Upper Volta (Burkina Faso) [26] and Tanzania [27] has shown that the dose required to reduce the fertility to 5% in treated males is 11 to 13 krad in air and 15 to 17 krad when the treatment is given in a protective nitrogen atmosphere.

Among the findings from these and other investigations are the following:
(i)   The eclosion rate of puparia in advanced developmental stage is not affected by the treatment.
(ii)   Irradiated males are fully competitive with untreated ones, and have unchanged abilities to form and transfer a spermatophore with motile sperm at normal frequency once they have reached sexual maturity (e.g. 5 to 7-day old G. p. palpalis males transfer 3 spermatophores within 24 hours).
(iii) Mating of treated males with untreated/wild female flies causes recurrent induced sterility in the fertilized egg. Embryonic arrest and zygotic death lead to physiological imbalance (e.g. building up of excess fat body) and aberrations in the reproductive system (e.g. prolonged retention of the degenerating embryonated egg in the uterus) [28].
(iv)   Multiple mating in young non-teneral nulliparous female flies is a frequently observed phenomenon but in most tsetse species does not result in significant changes in the level of induced sterility, providing the males received the recommended optimal treatment. Moreover, data from experiments in which the rate of remating and eventually multiple insemination by both sterile and fertile males was experimentally exagerated, indicate a strong tendency for precedence in usage of sperm from the first mating [29][30].
(v)   Female flies are completely sterlized when exposed to radiation doses above 5 krad (in air). After completion of oogenesis in the first two functional egg tubes and ovulation of the respective egg, ovary atrophy is initiated.
       During tests with G. p. palpalis, it was found that such treated females unlike untreated nulliparous and parous ones, do not show reluctance to re-mating.
       On the basis of dissection and microscopic examination, fairly simple procedures could be proposed for monitoring the effects and progress of sterile male releases in the field and for indirect detection of extremely low relic wild tsetse populations, particularly when the natural population has been reduced to levels below the threshold detectable by conventional techniques [29].

3. CONCLUSIONS

The ability to colonize, mass rear and optimally sterilize tsetse flies has increased enormously during the past decade. Some of the so-called "advantages, disadvantages and uncertainties" accredited to the Sterile Insect Technique and its use against tsetse flies have been more objectively approached and assessed by means of applied research under both laboratory and field conditions.

The success of pilot and operational field projects in Tanzania, Burkina Faso and Nigeria, justifies further efforts and investments for fine-tuning the SIT technology with the objective of integrating it with novel, environmentally safe methods to eradicate selected species of tsetse from various parts of Africa whenever it is economically and operationally feasible to do so. However, in considerating its future use especially in zones with multiple vectors of trypanosomiasis, certain areas of both field and laboratory research, essentially of an operational, physiological, biochemical and genetical nature need further attention in relation to their value for improving population detection, monitoring, suppression, mass-rearing and release methods.

As far as future research and development at the Seibersdorf laboratory and co-operating centres are concerned, priorities will be given to the following aspects:

(i)   Adaptation of G. pallidipes (Lambwe Valley strain) to the in vitro feeding system and introduction of other tsetse species in the laboratory (e.g. G. fuscipes, G. brevipalpis) with the aim of establishing productive autonomous colonies such as realized for G. p. palpalis.

(ii) Attempts to improve the nutritional quality of bovine blood through the use of inexpensive additives. The rationale behind this, is that bovine blood, is widely available in Africa as a waste product at slaughter-houses. In this context, optimization of the use of anticoagulants, phagostimulants and antimicrobial agents that are not harmfull to the flies is very important. Equally important are methods of storing blood to ensure availability in the process of mass-rearing.

(iii) Development of a cheap and practical synthetic diet and study of specific dietary factors which influence the cyclic development and transmission of trypanosomes in the target Glossina species.

(iv) Design and establishment of mass-rearing equipment and holding systems capable of ensuring a continuous monthly production of 500,000 puparia for at least two species of tsetse, and that can be operated economically and conveniently without sacrificing quality. Important in this respect is the need to determine: (a) the factors that control the optimal fly density in various cage designs (b) feeding attractants and stimuli enabling the use of large cages and the use of closed feeding systems (c) holding conditions which enhance synchronization of larvipositions and puparial eclosions.

REFERENCES

1. AZEVEDO, J. FRAGA DE. (1970). Prospects about international research work on tsetse flies. Proc. 1st Int. Symp. on Tsetse Breeding under Laboratory Conditions and its Practical Application. Lisbon, 22-23 April 1969, 497-502.

2. KNIPLING, E.F. (1963). Possible role of the Sterility Principle for tsetse fly eradication. Report WHO/Vector Control/27, 10 April 1963, 17 pp.

3. INTERNATIONAL ATOMIC ENERGY AGENCY. (1972). Application of sterility principle for tsetse fly suppression. At. Energy Rev. 10, 101-130

4. NASH, T.A.M., JORDAN, A.M., TREWERN, M.A. (1971). Mass-rearing of Tsetse Flies (Glossina spp.). Recent advances. Proc. Symp. on Sterility Principle for Insect control or Eradication, IAEA-SM-138/46, 14-18 Sept. 1970, 99-110.

5. Langley, P.A. (1970). Utilization of fat reserves and blood meals by tsetse flies in the laboratory: A comparison between Glossina morsitans Westwood and Glossina austeni Newstead. Proc. 1st Int. Symp. on Tsetse Breeding under Laboratory Conditions and its Practical Application. Lisbon, 22-23 April 1969, 265-271.

6. MEWS, A.R. BAUMGARTNER, H., LUGER, D., OFFORI, E.D. (1976). Colonization of _Glossina morsitans morsitans_ Westw. in the laboratory using _in vitro_ feeding techniques. Bull. Ent. Res. 65, 631–641.
7. WETZEL, H., BAUER, B. and BAUMGARTNER, H. (1974). The objectives of tsetse research in the Seibersdorf Laboratory (I.A.E.A.). Actes du Colloque sur Les Moyens de lutte contre les Trypanosomes et leurs Vecteurs, Paris 12–15 Mars 1974, 63.
8. BAUER, B., WETZEL, H. (1976). A new membrane for feeding _Glossina morsitans_ Westw. (Diptera, Glossinidae). Bull. Ent. Res. 65, 563–565.
9. OFFORI, E.D. and DORNER, P.A. (1975). Techniques for rearing _Glossina tachinoides_ Westw. Proc. Symp. on Sterility Principle for Insect Control, IAEA-SM-186/50, Innsbruck 22–26 July 1974, 487–494.
10. VAN DER VLOEDT, A.M.V. (1982). Recent advances in tsetse mass–rearing with particular reference to _Glossina palpalis palpalis_ (Rob.–Des,)fed _in vivo_ on guinea pigs. Proc. Symp. on Sterile Insect Technique and Radiation in Insect Control, IAEA-SM255/11, Neuherberg 29 June–3 July 1981, 223–252.
11. MEWS, A.R., LANGLEY, P.A., PIMLEY, R.W., FLOOD, M.E.T. (1977). Large scale rearing of tsetse flies (_Glossina_ spp.) in the absence of a living host. Bull. Ent. Res. 67, 119–128.
12. BAUER, B. and AIGNER, H. (1978). _In vitro_ maintenance of _Glossina p. palpalis_. Bull. Ent. Res. 68, 393–400.
13. BAUER, B., IWANNEK, K.H., HAMANN, H.J., ADAMSKY, G. (1980). Use of gamma–irradiated blood for feeding tsetse flies. Proc. Symp. on Isotope and Radiation Research on animal Diseases and their Vectors, IAEA-SM-240/13, Vienna 7–11 May, 1979, 319–326.
14. JAENSON, T.G.T. (1982). Radiosterilized, freeze–dried blood in the membrane-feeding for the tsetse fly _Glossina palpalis palpalis_. Ent. exp. & appl. 32, 281–285.
15. WETZEL, H. (1980). The use of freeze–dried blood in the membrane feeding of tsetse flies (_Glossina p. palpalis_, Diptera, Glossinidae). Tropenmed. Parasitol. 31, 259–274.
16. LUGER, D. (1982). Use of freeze–dried blood for mass–rearing of tsetse flies. Proc. Symp. on Sterile Insect Technique and Radiation in Insect Control, IAEA-SM-255/253, Neuherberg 29 June–3 July 1981, 217–221.
17. VAN DER VLOEDT, A. (1974). L'élevage au laboratoire de _Glossina palpalis palpalis_ et de _Glossina fuscipes quanzensis_, Actes du Colloque sur les Moyens de Lutte contre les Trypanosomes et leurs Vecteurs, Paris 12–15 Mars 1974, 61–62.
18. DELOACH, J.R. and TAHER, M. (1983). Investigations on development of an artificial diet for _in vitro_ rearing of _Glossina palpalis palpalis_. J. Econ. Entomol. 76, 1112–1117.
19. DELOACH, J.R. (1983). Sonication of reconstituted freeze–dried bovine and porcine blood. Effect on fecundity and puparial weight of the tsetse fly. Comp. Biochem. Physiol. 76A, 47–49.
20. KABAYO, J.P., DELOACH, J.R., SPATES, G.E., HOLMAN, G.M. and KAPATSA, G.M. (1986). Studies on the biochemical basis of the nutritional quality of tsetse fly diets. Comp. Biochem. Physiol. 83A, 1, 135–139.
21. KABAYO, J.P., TAHER, M. and VAN DER VLOEDT, A.M.V. (1985). Development of a synthetic diet for Glossina (Diptera: Glossinidae). Bull. Ent. Res. 75, 635–640.
22. ITARD, J. (1968). Stérilisation des mâles de _Glossina tachinoides_ aux rayons gamma. Rev. Elev. Méd. Vét. pays Trop. 21, 479–491.
23. CURTIS, C.A., LANGLEY, P.A. (1972). Use of nitrogen and chilling in the production of radiation induced sterility in the tsetse fly _Glossina morsitans_. Ent. exp. & appl. 15, 360–376.
24. LANGLEY, P.A., CURTIS, C.F., BRADY, J. (1974). The viability, fertility

and behaviour of tsetse flies (<u>Glossina morsitans</u>) sterilized by irradiation under various conditions. Ent. exp. & appl. 17, 97-111.

25. VAN DER VLOEDT, A.M.V., TAHER, M., and TENABE, S.O. (1978). Effects of gamms radiation on the tsetse fly, <u>Glossina palpalis</u> (Rob.-Desv,) with observations on the reproductive biology. Int. J. Appl. Radiat. Isot. 29, 713-716.

26. TAZE, Y., CUISANCE, D., POLITZAR, H., CLAIR, M., SELLIN, E. (1977). Essais de détermination de la dose optimale d'irradiation des mâles stériles dans le région de Bobo Dioulasso (Haute Volta). Rev. Elev. Méd. Vét. Pays Trop. 89, 269-279.

27. WILLIAMSON, D.L., BAUMGARTNER, H., MTUYA, H., GATES, D.B., COBB, P.E. and DAME, D.A. (1983). Integration of insect sterility and insecticides for control of <u>Glossina morsitans morsitans</u> Westwood (Diptera: Glossinidae) in Tanzania. II. Methods of sterilisation, transportation and release of sterilised males. Bull. Ent. Res. 73, 2, 267-273.

28. MATOLIN, S., VAN DER VLOEDT, A.M.V. (1982). Changes in the egg of the tsetse fly <u>Glossina palpalis palpalis</u> (Diptera: Glossinidae), after fertilization by sperm of gamma-irradiated males. Proc. Symp. on Sterile Insect Technique and Radiation in Insect Control, IAEA-SM-255/31, Neuherberg. 29 June-3 July 1981, 155-168.

29. VAN DER VLOEDT, A.M.V. and BARNOR, H. (1984). Effects of Ionizing Radiation on Tsetse Biology. Their relevance to entomological monitoring during integrated control programmes using the Sterile Insect Technique. Insect Sci. Applic. 5, 5, 431-437.

30. CURTIS, C.F. (1980). Factors affecting the efficiency of the sterile insect release method for tsetse. Proc. Symp. on Isotope and Radiation Research on their Vectors, IAEA-SM-240/25, Vienna 7-11 May, 1979, 381-394.

TABLE I - Gross data on the dynamics of the _Glossina palpalis palpalis_ (Seibersdorf) stock colony maintained on the membrane feeding system

	1981	1982	1983	1984	1985	TOTAL
No. of female flies[1]	2,502 - 24,133	8,622 - 24,078	10,460 - 30,800	25,200 - 70,700	47,800 - 72,300	
No. of puparia	236,362	446,284	509,650	1,148,306	1,404,635	3,745,237+[3]
Fecundity per month[1] [2]	2.00 - 2.57	2.00 - 2.51	2.04 - 2.97	2.00 - 2.96	2.37 - 3.06	
Mean value + S.D.	2.33 ± 0.19	2.23 ± 0.14	2.44 ± 0.29	2.58 ± 0.30	2.68 ± 0.21	
% daily mortality per month[1]	0.55 - 1.45	0.71 - 1.26	0.63 - 1.75	0.60 - 1.13	0.43 - 1.10	
Mean value + S.D.	0.84 ± 0.22	0.95 ± 0.16	1.08 ± 0.27	0.90 ± 0.56	0.63 ± 0.16	
No. of puparia to BICOT	29,000	50,000	74,200	325,000	405,800	884,000
Female flies emerged in Seibersdorf	99,545	139,460	201,641	279,448	397,522	1,117,616
Male flies emerged in Seibersdorf	71,142	97,766	136,262	175,043	280,751	760,964

1/ Minimum and maximum values during year

2/ Fecundity expressed as number of puparia per female >18 days (flies kept for 80 days)

3/ Ca. 311 000 puparia were discarded, 90 200 distributed to other centres, 240 850 used for local experiments and training, 149 320 under incubation

TABLE II – Performance during 1985 of all production units in the KM$_A$ colony of _Glossina palpalis palpalis_ fed 5 days a week on a 50:50 mixture of processed FPB + FBB

	Initial N ♀♀	Termination date 1985	day	Survivors %	Type of mortality (%) st	Bl	IUp	Total N puparia	A-class %	Fecundity P/initial ♀ Gross	Net
1984											
OCT 2	4925	09.01	91	51.4	40.6	4.6	3.7	21238	3.2	4.31	4.18
OCT 3	6525	16.01	88	61.5	30.5	4.3	3.7	28889	2.7	4.42	4.31
NOV 1	1500	23.01	84	50.7	39.3	7.0	2.8	6436	4.1	4.29	4.11
NOV 2	1275	30.01	81	63.8	26.2	6.5	3.4	5514	2.3	4.32	4.22
NOV 3	1750	13.02	85	64.8	22.7	8.4	4.0	7487	2.1	4.28	4.18
DEC 1	1700	20.02	82	55.2	38.2	4.4	2.1	7112	2.6	4.18	4.07
DEC 2	4350	06.03	86	61.9	31.4	4.5	2.0	19338	2.6	4.44	4.32
DEC 3	5925	13.03	83	64.2	30.2	3.2	2.3	24747	2.2	4.17	4.08
1985											
JAN 1	4900	27.03	86	57.4	37.0	3.1	2.3	20618	2.9	4.20	4.08
JAN 2	4725	03.04	83	54.2	39.7	3.6	2.4	19016	3.0	4.01	3.90
JAN 3	7069	10.04	80	52.5	41.1	4.0	2.3	27570	3.3	3.87	3.77
FEB 1	4325	24.04	83	50.4	44.0	2.9	2.5	17703	3.7	4.09	3.94
FEB 2	5705	08.05	87	52.6	41.1	2.7	3.5	25042	3.1	4.38	4.25
FEB 3*	2573	15.05	84	45.9	48.5	2.9	2.5	11105	3.2	4.30	4.16
MAR 1*	2520	22.05	83	52.2	41.7	3.5	2.6	10749	3.0	4.26	4.14
MAR 2*	3990	29.05	80	56.3	37.5	3.1	3.1	16067	2.9	4.02	3.91
MAR 3*	4950	12.06	84	64.2	29.5	3.3	2.8	21790	2.3	4.40	4.29
APR 1*	5040	26.06	87	59.8	33.8	4.0	2.3	22211	2.2	4.40	4.31
APR 2*	6930	03.07	84	70.9	23.0	4.1	1.9	30666	1.9	4.42	4.34
APR 3*	4560	10.07	81	64.6	29.2	4.5	1.6	19046	1.6	4.17	4.11

Cage density: 25 or 30* ♀♀ per standard cage (12.5 cm diameter)

st – Starvation   Bl – Blood in abdomen   IUp – In utero pupation

(KM$_A$ is one of the two existing stock colonies of _G.p.palpalis_ at the Seibersdorfer Laboratory)

TABLE II - (continued) Performance during 1985 of all production units in the KM A colony of *Glossina palpalis palpalis* fed 5 days a week on a 50:50 mixture of processed FPB + FBB

	Initial N ♀♀	Termination date 1985	day	Survivors %	Type of mortality (%) st	Bl	IUp	Total N puparia	A-class %	Fecundity P/initial ♀ Gross	Net
1985 (continued)											
MAY 1	5400	31.07	82	57.9	35.3	5.4	1.3	25257	1.9	4.00	3.94
MAY 2	6300	14.08	86	54.0	39.7	4.5	1.8	23680	1.7	3.98	3.92
MAY 3	5940	21.08	82	68.5	24.7	4.6	2.2	17049	0.8	4.17	4.16
JUN 1	4080	04.09	86	55.8	37.7	4.3	2.1	12861	1.3	4.37	4.32
JUN 2	2940	11.09	83	64.9	26.8	5.0	3.2	13070	1.0	4.27	4.23
JUN 3	3060	23.09	84	54.9	34.0	8.4	2.6	2647	1.2	4.64	4.58
JUL 1	8130	25.09	87	60.9	32.0	4.4	2.6	36814	1.0	4.52	4.48
JUL 2	5580	02.10	84	63.2	29.4	4.7	2.7	24561	1.0	4.40	4.36
JUL 3	5490	16.10	88	63.4	27.6	6.2	2.7	24745	0.9	4.50	4.47
AUG 1	570	23.10	84	54.9	34.0	8.4	2.6	2647	1.2	4.64	4.58
AUG 2	2790	06.11	88	62.5	25.4	8.5	3.5	12481	1.2	4.47	4.41
AUG 3	4620	13.11	85	67.5	22.9	6.9	2.6	19691	0.9	4.26	4.23
SEP 1	5850	27.11	88	62.8	28.6	5.7	2.8	27272	1.0	4.66	4.62
SEP 2	6750	04.12	85	65.6	26.7	5.2	2.4	29261	1.0	4.33	4.29
SEP 3	3300	18.12	89	52.8	34.5	9.0	3.6	14257	1.2	4.32	4.26

Additional production units formed: OCT 1: 5580; OCT 2: 6000; OCT 3: 6000;
NOV 1: 8100; NOV 2: 6000; NOV 3: 4980;
DEC 1: 6000; DEC 2: 6000; DEC 3: 8910;

st - Starvation   Bl - Blood in abdomen   IUp - In utero pupation

173

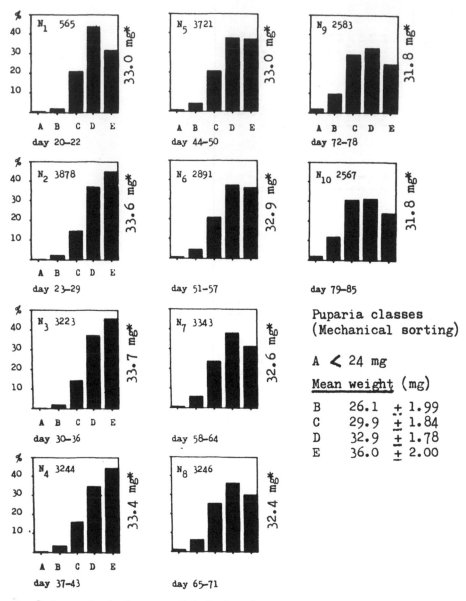

Fig. 1 Changes in the frequency distribution of puparial weights (5 classes) produced in *G. p. palpalis* membrane-fed stock colony units (Unit Sep. II, 1985: Initial N ♀♀ 6750; Total N puparia 29261)

* Total mean weight

Cage density: 30 mated ♀♀ per PVC cage (12.5 cm ⌀)

<u>Unit performance</u> (day 85)(5 feeding days a week)

Survival: 65.6%
Fecundity: 4.33 (gross) and <u>4.29</u> (nett: Class A puparia excluded) puparia per initial female

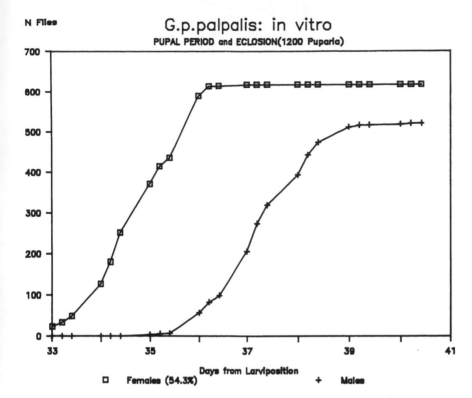

N Files

G.p.palpalis: in vitro
PUPAL PERIOD and ECLOSION(1200 Puparia)

Days from Larviposition

□ Females (54.3%)   +   Males

(*Three fly collections   day: 09.00   12.00 and 15.00 hr)

Incubation conditions: $23.5^{\circ} - 24.5^{\circ}C$ and $85 \pm 5\%$ R.H.

	Class B	Class C	Class D	Class E	Total
N puparia	300	300	300	300	1200
Emergence (%)	94.0	95.3	95.3	94.3	94.8
Female ratio	40.1	48.1	57.3	71.4	54.3
Mean pupal period (days)					
Females	34.35	34.83	35.07	35.19	34.92
Males	37.25	37.52	37.51	37.39	37.41

Fig. 2   Data on the eclosion rate and length of the pupal
period in the G. p. palpalis in vitro-fed stock colony
(random sample from the same larviposition-day).

The Antwerp tsetse fly laboratory: A new start

W.L.Decleir
Laboratory for Biochemistry and General Biology, RUCA, Antwerpen, Belgium

Summary

After the decease of professor Evens, the continuation of a
tsetse fly laboratory at the Antwerp State University Center
has been uncertain until two years ago, when the author decided to start a new research team for the study of the biology
of tsetse flies. This paper reports on the first results, obtained on the following subjects:

1. An automatic feeding device for rearing tsetse flies.

2. A tan eyed mutant colony of Glossina palpalis palpalis.

3. The possible effect of hormones in the diet on the fecundity of tsetse flies.

4. The effect of ivermectin on the reproduction biology of
Glossina palpalis palpalis.

1.   Introduction

Since more than twenty years the State University of Antwerp
has played an important roll in tsetse fly research with professor F.Evens and Dr. Van der Vloedt as principal investiga-
tors. After the decease of professor Evens, in 1982, the scientific staff of the laboratory disappeared from the university and the continuation of a tsetse fly research laboratory in
the RUCA became a matter of great concern. The author and his
colleague, professor W.Verheyen, both zoologists, managed to
save the existing tsetse fly cultures of Glossina palpalis palpalis (Zaire and Kaduna strains) and Glossina morsitans centralis. When finally, about 2 years ago, the author "inherited" the
laboratory of professor Evens with its infrastructure and know-
how for rearing tsetse flies, he decided to try a new start of
the Antwerp tsetse fly laboratory. With the stimulating help
and advice of professor W. Verheyen, director of the central
animalarium of the university center (RUCA), Dr. Van der Vloedt,
director of the tsetse fly laboratory at Seibersdorf and professor Mortelmans of the Institute for Tropical Medecine, he
started a new research group for the study of the biology of
tsetse flies and a research program with the available stock
colonies was established. This papers describes the very preliminary first results of this new research program.

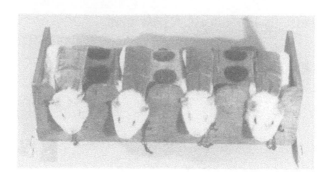

Photo 1

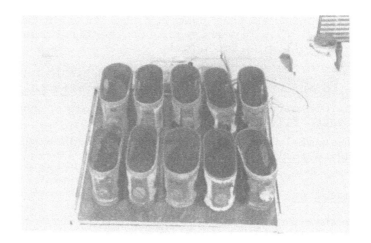

Photo 2

2. Description and results.

A. AN AUTOMATIC FEEDING DEVICE FOR REARING TSETSE FLIES.

Photo 1 shows the classical IN VIVO feeding system used in our laboratory with immobilized guinea pigs being exposed to cages with tsetse flies, covered with terylene netting.

Photo 2 shows our artificial feeding equipment using synthetic membranes, kept at 37-38°C. and mixed heparinized pig and bovine blood.

Photos 3 and 4 show our automatic feeding device, which, very recently, gave very good results for Glossina palpalis palpalis strains.
On photo 3 the aluminium U-profile, covered with a silicone membrane, is shown in an inclined plane. Photo 4 shows its actual position during the feeding of the flies.

178

Photo 3

Photo 4

1 = Multichannel peristaltic pump, assuring a continuous fluid flow of 26 ml/minute. Blood and rinsing solution are pumped from a refrigerator outside the rearing chamber.

2 = Aluminium U-profile (length 2 meters) with at the bottom 4 grooves (1 cm wide and 1 mm deep).

3 = Silicone membrane covering the 4 grooves.

4 = Tsetse fly cages on a PVC gutter.

5 = Heating tape (15 W/m. - 50 Hz - 12 Amp.max), fixed inside the U-profile together with a platinum film temperature sensor. The U-profile is filled up with 5 cm thick poly-styrene isolation to prevent loss of temperature at the upper side of the feeding unit.

179

6 = A series of 24 tsetse fly cages, each containing 30 producing females, can be fed simultaneously by 1 U-profile with
a length of 2 meters.

7 = Wiring leading outside the climate chambers to a temperature control unit and a timer for a 7 day multichannel program.

This automatic feeding system was invented and tested by Dr. F.
Bernaerts and Myriam Goossens,2 zoologists of our new tsetse
fly research team.

The warming up time (from 25°C. to 37°C.) is 30 minutes. Blood
input lasts 10 minutes and feeding time is 20 minutes. After -
wards the tubing and grooves are rinsed for 60 minutes with a
sodium citrate solution (10 mg/liter).
Each feeding unit (U-profile with a length of 2 meters) can be
used for feeding a series of 24 cages with each 30 flies. In
our climate rooms 6 feeding units can be mounted on each side.
This means that we foresee the automatic feeding of 12 X 24 X
30 = 8.640 flies in one rearing room. This system should be
operational in the course of 1986.

B. A TAN EYED MUTANT COLONY OF GLOSSINA PALPALIS PALPALIS.

In march 1983 the first technician of the Antwerp tsetse fly
laboratory, F. D'Haeseleer, isolated a mutant male with tan
coloured eyes in the Glossina palpalis palpalis stock colony,
originally derived from Kaduna puparia, supplied by Dr. Van
der Vloedt (I.A.E.A. - Vienna) and bred as an autonomous strain
since 1974. Six wild type females could be inseminated before
the mutant died and were the origin of a mutant stock colony.
Breeding experiments indicated that the mutation is a sex-link-
ed recessive trait, which resides at the differential part of
the X-chromosome. The maintenance of the mutant stock does not
require any special precautions. Actually the laboratory keeps
a tan eyed colony of approximately 2.000 producing females and
puparia are available for eventual distribution to other laboratories.

C. THE POSSIBLE EFFECT OF HORMONES IN THE DIET ON THE FECUNDI-
TY OF TSETSE FLIES.

Very recently Myriam Goossens and Andrea Vlaeminck started a
physiological study of tsetse flies. In a first experiment
they fed female and male tsetse flies (Glossina palpalis palpalis - Zaire strain) on female, male and castrated male guinea
pigs and found that the flies produced less puparia, as compared with control animals, after a test period of 50 days, when
both partners were fed on castrated host animals. The survival
and puparial weight however were not affected. These results
point to a possible effect of hormones in the diet on the fecundity of tsetse flies. However more research is needed to
explain these results.

D. THE EFFECT OF IVERMECTIN ON THE REPRODUCTION BIOLOGY OF
GLOSSINA PALPALIS PALPALIS.

One of the control methods of tsetse fly populations is the
use of antiparasitic drugs in the host animals. One of these

is ivermectin, a commercially available antihelminthic agent which also acts against a series of arthropod ecto- and endo-parasites of domestic animals (1).
The activity of ivermectin against <u>Glossina</u> sp. was first reported by Distelmans et al. (2), who worked in the RUCA tsetse fly laboratory. Later on Langley et al. (3) showed that the effect of ivermectin on female tsetse flies consists of a reduction of fecundity by inducing abortion. In 1985 the study of the effect of ivermectin on the reproductive biology of <u>Glossina palpalis palpalis</u> was started again in the RUCA-laboratory due to the stimulating advice of professor Mortelmans. J. Van den Abbeele, M. Goossens and F. D'Haeseleer and recently also E. Verheyen injected rabbits with different doses of ivermectin (0.2 mg /kg, 0.4 mg/kg, 1.0 mg/kg and 2.0 mg/kg), used them for feeding female flies and studied their effect on mortality, puparia formation , insemination status(by examination of the spermathecae) and ovary configuration (4). They found no significant effect of the recommended therapeutic dose (0.2 mg/kg) on reproduction. Higher doses however caused a significant reduction of reproductive performance and they could explain this by a retardation of the ovulation/pregnancy cycle and a disturbance of the pupariation process in several larvae. They, however, found no abnormal abortion rates and no significant difference of emergence rates and sex-ratios in the offspring of treated and control series.

REFERENCES.

1. CAMPBELL W.C. and BENZ G.W. (1984). Ivermectin: A review of efficacy and safety. J.Vet.pharmacol.Therap. 7,pp 1-16.
2. DISTELMANS W., D'HAESELEER F.,MORTELMANS J. (1983). Efficacy of systemic administration of ivermectin against tsetse flies. Ann.Soc.Belge Méd. Trop. 63, pp. 119-125.
3. LANGLEY P.A. and ROE J.M. (1984). Ivermectin as a possible control agent for tsetse fly,<u>Glossina</u> <u>morsitans</u>. Entomol. exp. appl. 36,pp. 137-143.
4. VAN DEN ABBEELE J.,GOOSSENS M. and D'HAESELEER F. (1986). Efficacy of ivermectin on the reproductive biology of <u>Glossina</u> <u>palpalis</u> <u>palpalis</u>. Ann. Soc.Belge Méd. Trop.: In press.

Let us live with tsetse flies

J.Hardouin

Department of Animal Production and Health Institute of Tropical Medicine, Antwerp, Belgium

Summary

Chemotherapy and prophylaxis can cope with temporary risks linked
with animal trypanosomiasis. It would be a mistake to undertake
campaigns to allow introduction of zebu cattle in presently tse-tse
flies infested areas. On the contrary, trypanotolerant cattle should
be favoured in such areas as it is now proved that they are more
productive through their light body weight than the larger zebus.
In this context the dwarf Guinean goat and Djallonke sheep deserve
also more consideration, as well as the potentialities offered by
rational use of wild animals in a given situation. Tse-tse eradication programmes are not trustworthy as sustained success is extremely
rare. Simple and cheap techniques to control the fly population should
be further studied and used extensively. More financial resources
should be made available for investigation on the mechanisms of
trypanotolerance, and to multiplication of trypanotolerant breeds if
required. A fair and unbiased analysis of each situation should be
undertaken on merit basis without any preconceived ideas with due
consideration on the conservation of the environnement.

1. Introduction

The importance of animal trypanosomiasis is well known and there is
hardly a need to further emphasise the ravaging effects of the disease
and everyone agrees to it. However, it is not the same concerning the
attitude to be adopted when one is faced with the problem of sleeping
sickness, as diverse possibilities exist to achieve the goal. A general
trend is to give extreme importance to methods geared towards control of
the tse-tse flies, though it is not proven if these are the best strategies. The aim of this paper is to plead for much more attention to the
animals and more funds needed for research on trypanotolerance.
In some cases the drugs do suffice when animals harbouring trypanosomes are treated. It can be applied on animals which have to be transfered
through or have to remain in infested areas, such as during the movement
to slaughterhouses or in feed lots for fattening during a short period.
Prophylactic and therapeutic drugs are available against trypanosomiasis
which are rather well known and used by veterinarians and animal production specialists. Nothing more is required.
A joint F.A.O./W.H.O. mission in Africa showed me recently a rather
very different attitude of the specialists dealing with human trypanosomiasis and those dealing with livestock trypanosomiasis. The W.H.O.partici-
pants clearly stated indeed that, at least for the concerned areas, the
fight against the flies was not interesting since treatments of people
showing signs of human sleeping sickness gave them satisfaction. In such
instances there is no need for a sophisticated cost/benefit analysis to

justify for such an attitude as the number of involved annual cases is just
few dozens or hundreds. Moreover, there are now only few areas where human sleeping sickness is the real limiting factor for development, especially when we take into consideration the fact that extension of agriculture is probably the most adequate method to eliminate the tse-tse
flies.

2. The Trypanotolerant Breeds

The argument to continue and extend campaign programmes against tsetse flies is based on the assumption that the concerned areas have to be
developed through extensive rearing of zebu cattle, which in view of the
recent studies is a very short-sighted approach. It is now proved that
the productivity of the trypanotolerant cattle is higher than that of the
zebu cattle, and the lighter body weight of the taurines is an advantage
as is shown by the F.A.O./I.L.C.A. productivity index, by the yard-
stick of the Relative Daily Weight Gain and other considerations.
The introduction of zebu cattle in an area where this animal cannot sur-
vive under existing situations means that the whole biotope has to be
changed. One has to keep in mind of all its possible consequences such as
on the climate, rainfall, soil fertility, firewood resources,... It is
indeed extremely surprising that the choice is still under consideration
between not modifying anything in the environment and favour the rearing
of adapted trypanotolerant breeds,or to manipulate the environment and
introduce unadapted foreign breeds of animals. In other words, one has to
choose between no modification of the ecosystem or modify the two components of the ecosystem.
It is often said that there are no trypanotolerant cattle available
to stock new areas , but this again is an aphorism which nobody has seriously tried to check. There are, however, evidences that the supply of trypanotolerant cattle can cope with this potential demand, as presently investigated by the F.A.O. project GCP/RAF/190/ITA " Amélioration, Multiplication et Conservation du Bétail Trypanotolérant en Afrique de l'Ouest".
Moreover, no one is prepared to ascertain with confidence that action has
really been taken on a regional basis to built up multiplication farms and
salvage all the available animals. Simple calculations have shown that,
with adequate production parameters, a cattle herd left for breeding in
these areas multiplies several times in number within a few years.
Recent studies have also shown that trypanotolerant strains also
exist in the zebu population, but very little effort if at all appears
to have been made to investigate this aspect in depth and to multiply
such trypanotolerant zebus. Generally speaking, the moral or financial
support for adequate research on the trypanotolerance mechanisms is too
scarce.
It seems also necessary to remind the planners that the local Djallonke sheep and the local dwarf Guinean goat are perfectly adapted to hot
and humid environments and that their level of trypanotolerance seems to
be higher that those of cattle. Cases where trypanosomes have been seen
in blood smears from these small ruminants are extremely rare, which is
not the case for the Ndama and other West African taurine breeds of
cattle. Luckily enough , the international organizations have now recognized the tremendous importance of sheep and goats in rural development as
they are of direct benefit for the peasants with marginal farming system
for whom the social role of the small ruminants has been overlooked until
recently. The demand for mutton and goat meat is often higher than for
beef in national meat markets of some of these areas. I would also like to

draw the attention of the readers of the under-exploited meat production
potentialities offered by many wild animal species which are living in
equilibrium with the tse-tse flies. Buffaloes, zebras, various antelopes,
warthogs do produce and reproduce in fly infested areas but most of the
veterinarians and zootechnicians lack the insight, follow a traditional
approach and adhere too strictly to the education princinles of our grand-
fathers' era. There is no competition between the exploitation of domestic livestock and wild animals as such in their natural environment, but it
should be a mistake to substitute livestock by game everywhere. Higher
profitability through lower expenditures have been noticed in many places
where difficult cattle ranching has been replaced by game or mixed livestock and game ranching.

A better use of any type of land is always obtained when many ecological niches are exploited, which is indeed the case with multi-species game
ranching or combined livestock and game ranching.

Marginal lands, like those still infested by tse-tse flies and trypanosomiasis, deserve more attention and a too narrow view point for the
control and/or the elimination of the parasites is farfetched.

3. Eradication Campaigns

I have not tried to evaluate the cumulative expenditures involved in
tse-tse flies eradication campaigns, but the amount is surely astonishing-
ly high and frightful. Excepting in very peculiar situations, the chance
of permanent success of any eradication campaign is extremely bleak.
The only cases where a sustained eradication programme can be implemented
with some confidence are those of oceanic islands, where successes have
indeed been achieved e.g. for Anopheles. Even the oases in
the middle of the Sahara, which are in effect vegetation islands in stone
or sand oceans, are regularly re-invaded by mosquitoes and malaria. Has
the yellow fever been definitively eliminated anywhere by the eradication
of its vector making the value of vaccination questionable ?
The readers would have understood that I do not support the notions
of fly eradication or control campaigns with the final aim of achieving
increased meat and milk production, but this does not mean that I should
discard the usefulness and validity of simpler and cheaper control measures
for limiting flies populations such as screens , tyre traps,... These
techniques should in fact be tested more extensively, controlled for their
value, recommended if adequate and applied on a wider scale. The fight
against animal trypanosomiasis can indeed give hopes for a possible vaccine. It appears to me that too much of financial support has already been
provided for the study of tse-tse fly biology at the cost of support which
is genuinely needed for certain basic investigations on the phenomena of
trypanotolerance and on the multiplication and installation of trypanotolerant livestock in tse-tse infested areas . Financial resources are limited
and a fair and objective choice ought to be made just now.

4. Conclusions

By virtue of me having worked and travelled for nearly thirty years
in Africa, and having been last month in West Africa for F.A.O. consultancy on adequate development in trypanosomiasis infected area, my position
is very clear : a global approach is required. Preconceived ideas are not
a scientific attitude, and all the elements and possible solutions for a
specific problem have to be analysed and weighted on their own merits.
As an animal production man I am by vocation, I would never dare to

recommend without hesitation heavy investments and recurrent costs for commercial livestock enterprises in areas where a tse-tse flies eradication campaign is under way, as I shall never be certain if reinvasion of flies will not occur. This is based on my animal health education and my own experience. I do hope many colleagues will be as realistic as I have tried for myself. Accept a moment of wishful thinking of your own money being involved in such eradication campaigns, as a financial investment, and you draw honestly the conclusion. It is my ironical thought that it would be much easier a task to eliminate elands, buffaloes or eliphants than the tse-tse flies.

Nowadays when environmental preoccupation takes more and more importance, as it should be, attempts to modify biotopes must be scrutinized beforehand and studies on its impact should be undertaken . There is never only one possible solution for a given problem. Whenever a choice has to be made, preference should be given to methods which do not modify the environment and make use of reciprocal equilibrium between flora, domestic and wild fauna, soil, climate, people,...

The general framework of the animal production development in presently tse-tse flies infested areas gives the chance of making the good choice. Let us try avoiding erroneous goal. The main target should not be the tse-tse fly but the producing animals. Let us learn to live with the tse-tse flies. Man kind dispose of many means to produce in difficult environments. Let us continue and increase the research on the mechanisms of the trypanotolerance in order to scientifically base the multiplication of breeds or strains on quantitative parameters of the trypanotolerance. The results already obtained in some years only is the more eloquent argument for continuation of this promising way.

Closing session

Chairman: C.Pelerents
Sessions' report by: A.M.Jordan
I.J.Everts
A.Dame
K.B.David-West
Conclusions by: C.Pelerents

Sessions' report

Introduction
A.M. Jordan

A review of present trypanosomiasis control methods with recommendations for future development of control strategies has been made.
The importance of integrating the controlled use of trypanocidal drugs with vector control methods was emphasised.
Before embarking on any campaign to control African trypanosomiasis it is essential to carefully define objectives as appropriate strategies will vary from one situation to another. In particular, is the human or animal disease being combatted and is the objective to eradicate the vector from a prescribed area with no prospect of reinvasion or is it only possible to control the disease by drugs or by reducing the numbers of vectors? The importance of linking control or eradication campaigns with detailed land use studies was emphasised.

Whereas insecticides will continue to be used in large-area eradication campaigns it is likely that because of high costs their use in regularly repeated control operations will become less widespread.

The recent development of insecticide-impregnated targets and associated odour baits offers the prospect of an effective, non-polluting method of achieving significant reductions of Tse-tse populations. The costs of deploying such devices over large areas should be assessed.

The range of techniques available, competently applied, can achieve control of African trypanosomiasis, in many circumstances at an acceptable cost. The limitations to the effective use of these methods are more logistic than technical, in particular the shortage of skilled manpower to plan and conduct operations in the field.

Safety and efficaciousness of chemical control
J.W. Everts

Tse-tse flies chemical control, in the past, often gave rise to considerable and sometimes irreversible damage in the ecosystems

189

concerned. The most dramatic side-effects were observed in situations where dieldrin and endosulfan were used at high dose rates.

The methods developped over the last decade for non-residual applications of certain pyrethroids and endosulfan cause less environmental damage and in most situations recovery of affected populations of organism can be observed within the same season.

Nevertheless, a continuing surveillance of possible environmental effects of pesticides in connection to Tse-tse control operations remains necessary.
One may expect that the resilience of many ecosystems may decrease in future because of factors related to the steady increase of human pressure on the environment.
Pesticide applications, whose side-effects appear to be of a reversible nature in relatively unspoiled habitats may cause irreversible damage in habitats, which, to some degree, have lost their recovering strenght.

An overview has been presented about environmental effects caused by pesticides used in Tse-tse control operations in Africa.
Only those insecticides was considered which have been applied at an operational level and whose side-effects were studied under field conditions. These are DDT, dieldrin, endosulfan, deltamethrin and permethrin.
Known data on side-effects of Tse-tse control operations have been presented, summarized in a condensed form.

Biological, biotechnical and other control methods
D.A. Dame

Although the use of classical bicontrol agents, such as parasites, pathogens and predators, cannot currently be considered for Tse-tse control, there are some very promising new biotechnical methods. The usefulness of attractant devices and sterile insects has been field-tested and both methods have been found to be extremely effective.

Attractant devices treated with pyrethroid insecticides and placed in riverine habitats at 100 to 200 meter intervals, or occasionally 20 to 30 meter intervals where the vegetation is very dense or in non-riverine habitats at the rate of 4 per square kilometer, are effective in all seasons and have provided control in excess of 90%. Against savannah species these traps, targets and screens have reduced fly populations below the detectable level. Enhanced by volatile, odor attractants, e.g. carbon dioxide, acetone, octenol, or animal byproducts, such as urine, these devices are more effective than standard surveillance methods for the collection of Tse-tse flies in both types of habitats. Their potential application as sites for contact sterilization of natural fly populations has been demonstrated in the field; further studies are currently being conducted to discover simplified methods to expose fly populations to chemosterilants without risk of exposure of other organisms to these mutagenic materials. The potential use of the attractant devices for collection of flies for manual "in situ" sterilization and release was discussed as a complementary method to supplement the conventional sterile insect technique of releasing factory-produced insects. Population simulations have revealed that relatively low daily collections (ca. 1%) of indigenous flies, which are then released as sterile insects, will cause population decline and bring about complete control more rapidly than insecticidal control through attraction at the same daily rate. Studies with combinations of chemosterilant and natural or synthesized Tse-tse pheromones are currently being conducted in the field to test the potential of this approach with simple attractant devices.

Relatively large scale production and release of sterile adults of the riverine species, <u>Glossina palpalis</u> <u>palpalis</u>, <u>G. palpalis</u> <u>gambiensis</u> and <u>G. tachinoides</u> have resulted in complete control, following prior population suppression with attractant devices treated with deltamethrin. Ratios of 10 sterile males to each indigenous, fertile male were required in the riverine situations. Releases of sterile savannah species, <u>G. morsitans morsitans</u>, in the pupal stage resulted in the theoretically expected reproductive decline at ratios at or below one sterile male to each indigenous fertile male. The riverine Tse-tse, reared on live host animals or on "in vitro" systems utilizing artificial membranes and bovine or porcine blood from local abattoirs were equally effective in the field. The future use of propagation facilities and their potential for providing sterile insects for other programs was discussed; there was consensus that the capability of these facilities should not be lost.

The role of computer modeling for population simulation was discussed and the need for software compatible with microcomputer equipment available to operational programs was emphasized. It is evident that computer models have already played a significant role in Tse-tse research; availability of similar user-friendly computerization for operational use is highly desirable.

The advisability of assigning a dedicated research unit to each major operational program was emphasized as an important factor related to the continuing need to cope with changing circumstances and maintain the capacity for program improvement. Methods using aerial or ground insecticide applications were seen to be compatible with attractant and sterility methods. In certain circumstances, it is possible that each method could be used alone, and in other circumstances it would be an advantage to integrate appropriate available methods. Each program must be considered separately to determine the most suitable strategy. A debate on the relative merit of control strategies, vs.

191

eradication strategies, revealed the need to carefully consider the relative merits of each approach. The prospect of living with human and animal trypanosomiasis indefinitely would necessitate the continuance of both livestock and human health services for this group of parasites. With over 10 million square kilometers of Glossina infestation, it will be necessary to make major decisions at governmental levels on which locations have characteristics that are suitable for eradication strategies and justify the expense involved. These decisions will be balanced by the costs of the alternative decision, i.e., to rely on continued fly suppression measures or to live with the existing fly populations and provide the necessary health and veterinary services.

Present and future programmes of
international and national organizations
K.B. David-West

The programmes and discussions were focused on the following:

1. Aerial spraying involving research on aerosol characteristics and field control programmes.
2. Use of traps and impregnated screens involving research on odour attractants, pheromone baited sterilising traps.
3. Sterile Insect Technique - emphasis was on cost effectiveness through improvement of mass rearing technique. There are future prospects for commercialisation of insects rearing for field control programmes. Pilot integrated SIT programmes have been succesful in Nigeria and Burkina Faso. There was some concern about the future of the Bobo Diulassa breeding project which will terminate this year. The possibility of its use as a regional breeding centre for the supply of pupae to other sub-units was discussed.
4. Chemotherapy - there is little research on chemotherapy and some concern was expressed on the termination of the screening of chemotherapeutics by GTZ in 1987. The discussions stressed the need for more investments in chemotherapeutic studies particularly for human trypanosomiasis.
5. Monitoring of environmental pollution as an integral part of any control programme using insecticide, was emphasised.
6. The use of bait animals with toxicants was discussed. References were made to Ivermectin and Delmathazine.
7. FAO programmes - it was observed that FAO has deviated from its original planned continental campaign but it has established a Tse-tse and Trypanosomiasis unit at its Headquarters and a Commission on African Trypanosomiasis and related matters.
8. Research on Trypanotolerance and Tse-tse challenge is being covered by the International Trypanotolerant Centre in Gambia and the ILCA/ILRAD Trypanotolerant network programme.

9. The need for community participation in Tse-tse control programmes at village level was emphasised. Some participants suggested that sociological studies on how to introduce such programmes at village level should be carried out.

For future programmes the following were considered:

I.   Aerial spraying - toxicity and formulation research
II.   Traps and screens - odor attractant, chemosterilisation and design
III.   Sterile Insect Technique - improvement of rearing technique
IV.   Pest resistance to insecticide
V.   Use of wild Tse-tse population for sterilisation and release in field (work on mobile sterilising unit is on-going)
VI.   Training - also communication
VII.   European Association of Tse-tse and Trypanosomiasis workers
VIII. Chemotherapy screening/drug resistance
IX.   Tse-tse biology and behavioural research which is oriented towards field control strategy
X.   Trypanotolerance/immunology
XI.   Development of vaccine against trypanosomiasis

Conclusions

C.Pelerents

The experts would like to express their thanks to the Commission of the European Communities for giving them the opportunity to participate in this Symposium.

Facts

- African trypanosomiasis is still a major problem which needs to be urgently resolved for the benefit of human health and the development of stock farming in vast regions. If the introduction of new non-toxic trypanocides and possibly a vaccine is essential for human trypanosomiasis, vector (Tse-tse flies) control occupies an important place among control methods, particularly in the field of veterinary medicine.

- Campaigns against Tse-tse flies have made progress but numerous problems still exist, calling for continued action of long duration.

- The joint effort by national and international organizations for Tse-tse fly control is well regarded and it is desirable that this be pursued further and, if possible, expanded.

- Several approaches exist for trypanosomiasis control and, more particularly, for vector control, which correspond to different techniques and to the specific conditions of the different agroecosystems concerned.
  None of these approaches, taken separately, seem able to lead to an economic and lasting control of the vector. It is thus indispensable to encourage the implementation of concrete integrated control strategies combining two or more techniques.

Recommendations

- The exchange of information between researchers, nations, national and international organizations, should be improved.
  The governments of the countries involved in the control projects should collaborate in sensitizing and involving the concerned populations by localized and adapted informative actions.

- Training courses for technicians and graduates should be continued and possibly intensified. Manuals for participants at all levels of the Tse-tse fly control projects should be made available.

195

- Resources should be placed at the disposal of well-established and well-structured national unities in order that they may continue to operate as regional, logistic centres.
An appropriate allocation of resources should take into account the diversity of control strategies to be applied to different species of Tse-tse flies in different agro-ecosystems.

- The control programmes should be preceded by a study of future landuse and the utilization of soils, followed by a long-term evaluation of the results of the programme (control efficiency,cost/benefits, environmental impact).

- If the insecticide treatments remain preponderant in Tse-tse fly control, these methods should still be improved. A particular effort should be made in the field of aerial-spraying on uneven terrain, the impact of products on the environment, the duration of action, the efficiency of insecticide, and the research of new products. The possibility of resistance of Tse-tse flies to these products should be monitored very closely and studied by specific method (e.g. genetic and molecular chemistry).

- The use of the Sterile Insect Technique (SIT) method should be reinforced at all levels, particularly rearing technique which is an essential condition for its application.

- The control methods using traps and attractants should be improved by "orientated" research into endocrinology and ethology of the vectors in order to better understand the mechanisms of reproduction, and the behaviour vis-à-vis traps, screens and attractants. The utilization of pathogene agents for the vector could also be envisaged.

- Methods and strategies should be set up and experimented with for the establishment of durable and efficient barriers against reinfesting, which is a frequent cause of long term setbacks in control campaigns.

- The study of the dynamic of the different wild populations of the vector (model) should be encouraged, as well as the role of wild fauna in the transmission of trypanosomiasis.

- In parallel with the research on vector control, new pharmaceutical products should be introduced and more particularly, the utilization of contact and systemic products against Tse-tse flies. Field trials should be undertaken in relation to the results obtained in the laboratory.

- Finally, the choice of bovine, caprine and ovine breeds to be raised in these regions should be based not only on their capacity of production but also on their level of trypanotolerance, the mechanism of which should be studied.

List of participants

Belgium:

       DECLEIR Walter
       Laboratory of Biochemistry and
       General Zoology - R.U.C.A.
       Groenenborgerlaan, 171
       2020 - Antwerpen

       GEERTS Stanny
       Institute of Tropical Medicine
       Nationale Straat, 155
       2000 - Antwerpen

       HARDOUIN Jacques
       Institute of Tropical Medicine
       Tropical Animal Production Unit
       Nationale Straat, 155
       2000 - Antwerpen

       MORTELMANS Jos
       Institute of Tropical Medicine
       Nationale Straat, 155
       2000 - Antwerpen

       PELERENTS Christian
       State University of Gent
       Coupure Links, 653
       9000 - Gent

Bundesrepublik Deutschland:

       BRUECKLE Fritz W.
       G.T.Z.
       Dag-Hammarskjoeld-Weg, 1
       P.O. Box 5180
       6236 - Eschborn

Burkina Faso:

       KABORE Idrissa
       C.R.T.A./G.T.Z.
       B.P. 454
       Bobo Dioulassa

France:

CLAIR Michel
I.E.M.V.T.
10, Rue Pierre Curie
94704 - Maisons Alfort

CUISANCE Dominique
I.E.M.V.T. - C.I.R.A.D.
10, rue Pierre Curie
94704 - Maisons Alfort

TACHER Georges
I.E.M.V.T.
10, Rue Pierre Curie
94704 - Maisons Alfort

Kenya:

MUTINGA Mutuku J.
International Centre of Insect
Physiology and Ecology
P.O. Box 30772
Nairobi

TURNER David
International Centre of Insect
Physiology and Ecology
P.O. Box 30772
Nairobi

Netherlands:

EVERTS James W.
Department of Toxicology
Agricultural University
De Dreyen, 12
6703 Wageningen

Nigeria:

DAVID-WEST Kelsey
Federal Ministry of Agriculture
PMB 12613
Lagos

Portugal:

PINHAO Rui Costa
Instituto de Higiene e
Medicina Tropical
Rua da Junqueira, 96
1300 - Lisboa

United Kingdom:

       ALSOP Nicholas John
       Carlton, Meadow Lane, Houghton
       Huntingdon

       FREELAND Guy
       Overseas Development
       Administration
       Eland House - Stag Place
       London

       JONES Tecwyn
       Tropical Development &
       Research Institute - O.D.A.
       College House - Wright's Lane
       London SW8 5SJ

       JORDAN Anthony Michael
       O.D.A./University of Bristol
       Langford House
       Bristol BS18 7DU

U.S.A.:

       DAME David A.
       Agricultural Research Service
       United States Department of Agriculture
       P.O. Box 14565
       Gainesville, Florida 32604

       TAKKEN Willem
       P.O. Box 14565
       1600 SW 23rd Drive
       Gainesville, Florida 32604

Zambia:

       LOHER Erhard
       Delegation of CEC
       Lusaka-Zambia
       P.O. Box 34871
       Lusaka

Zimbabwe:

       DE VRIES Jan
       Delagation of CEC
       Harare - Zimbabwe
       P.O. Box 4252
       Harare

HURSEY Brian
TseTse & Trypanosomiasis Control Branch
P.O. Box 8283 Causeway
Harare

LOVEMORE Desmond F.
Regional TseTse & Trypanosomiasis
Control Programme
P.O. Box A560 Avondale
Harare

VALE Glyndwr Alan
TseTse & Trypanosomiasis
Control Branch
Department of Veterinary Service
P.O. Box 8283 Causeway
Harare

INTERNATIONAL ORGANIZATIONS

C.E.C.

BISHOP George Robert
Commission of the European Communities
Joint Research Centre
I - 21020 Ispra

CAVALLORO Raffaele
Commission of the European Communities
Joint Research Centre
I - 21020 Ispra

DAVIES David Hywel
Commission of the European Communities
D.G. XII
200, rue de la Loi
B - 1049 Bruxelles

DE BRUYCKER Marc
Commission of the European Communities
D.G. XII
200, rue de la Loi
B - 1049 Bruxelles

DE VISSCHER Marie Noel
Commision of the European Communities
D.G. XII
200, rue de la Loi
B - 1049 Bruxelles

MULDER Jan
Commission of the European Communities
200, rue de la Loi
B - 1049 Bruxelles

UZUREAU Claude
Commission of the European Communities
D.G. XII
200, rue de la Loi
B - 1049 Bruxelles

F.A.O.

LE ROUX Jan G.
F.A.O.
Via delle Terme di Caracalla
I - 00010 Roma

I.A.E.A.

LINDQUIST D.A.
I.A.E.A.
Wagamerstrasse, 5
P.O. Box 100
A - 1400 Vienna

VAN DER VLOEDT A.M.V.
Joint F.A.O./I.A.E.A. Division
Insect and Pest Control Section
Wagramerstrasse, 5 - P.O. Box 100
A - 1400 Vienna

W.H.O.

DE RAADT Pieter
Parasitic Diseases Programme
World Health Organization
Via Appia
CH - 1200 Genève

List of authors

ALLSOPP, R. 123

BAUER, B. 71
BRÜCKLE, F.W. 103

DAME, D.A. 93, 190
DAVID-WEST, K.B. 153, 192
DAVIES, D.H. 3
DE BRUYCKER, M. 157
DECLEIR, W.L. 177
DE RAADT, P. 147

EVERTS, J.W. 49, 189

GEERTS, S. 43
GOLDER, T.K. 17

HARDOUIN, J. 183

JONES, T. 123
JORDAN, A.M. 7, 189

KABAYO, J.P. 163
KABORE, I. 71
KOEMAN, J.H. 49

LAVEISSIÈRE, C. 67
LE ROUX, J.G. 133
LINDQUIST, D.A. 139
LOVEMORE, D.F. 27
LUGER, D. 163

MORTELMANS, J. 43
MUTUKU MUTINGA, J. 113

OFFORI, E.D. 163

PELERENTS, C. 195
PERFECT, T.J. 123
PINHÃO, R.C. 161

TACHER, G. 117
TAKKEN, W. 83
TRAVASSOS SANTOS DIAS, J.A. 75
TURNER, D.A. 17

UZUREAU, C. 157

VALE, G.A. 59
VAN DER VLOEDT, A.M.V. 163